AF581700

Steady-State Operation, Disturbed Operation and Protection of Power Networks

Steady-State Operation, Disturbed Operation and Protection of Power Networks

Editor

François Vallée

MDPI • Basel • Beijing • Wuhan • Barcelona • Belgrade • Manchester • Tokyo • Cluj • Tianjin

Editor
François Vallée
Power Electrical Engineering Unit,
University of Mons
Belgium

Editorial Office
MDPI
St. Alban-Anlage 66
4052 Basel, Switzerland

This is a reprint of articles from the Special Issue published online in the open access journal *Energies* (ISSN 1996-1073) (available at: https://www.mdpi.com/journal/energies/special_issues/SDP_PN).

For citation purposes, cite each article independently as indicated on the article page online and as indicated below:

LastName, A.A.; LastName, B.B.; LastName, C.C. Article Title. *Journal Name* **Year**, *Volume Number*, Page Range.

ISBN 978-3-0365-0320-2 (Hbk)
ISBN 978-3-0365-0321-9 (PDF)

Contents

About the Editor

François Vallée received his degree in Electrical Engineering and a Ph.D. degree in Electrical Engineering from the Faculty of Engineering, University of Mons, Belgium, in 2003 and 2009, respectively. He is currently an associate professor and leader of the "Power Systems and Markets Research Group" at the University of Mons. His Ph.D. work was awarded by the SRBE/KBVE Robert Sinave Award in 2010. His research interests include PV and wind generation modeling for electrical system reliability studies in the presence of dispersed generation. He is currently a member of the Governing Board from the Belgian Royal Society for Electricians—SRBE/KBVE (2017)—and an associate editor of the International Transactions on Electrical Energy Systems (Wiley).

Preface to "Steady-State Operation, Disturbed Operation and Protection of Power Networks"

With the ongoing energy transition, distributed energy resources (DERs) and new loads (e.g., electric vehicles, EVs) are emerging in modern power systems, which highly impacts the operation of the latter. Indeed, in addition to an increased uncertainty in power system management, DERs as well as EVs can significantly affect the power quality level and contribute in multiple manners to faulty currents. Many algorithms and tools have been developed in recent years to ensure the safe operation of the system while fostering the integration of renewable energy-based generation. Moreover, the current advances in artificial intelligence and computation resources offer new prospects for related research.

This Special Issue offers a wide panel of up-to-date research that aims to ensure the enhanced operation of the electricity system. The first three contributions investigate how artificial intelligence (both machine and reinforcement learning) and new local business models (e.g., renewable energy communities) can assist in an effective management of modern distribution systems. In addition, it is essential to have an accurate knowledge of the network parameters when analyzing the stability and power quality of a power system. In this way, the network impedance versus frequency characteristic is a key element to monitor, involving a huge amount of data to be processed. Hence, the fourth paper of this Special Issue investigates how lossy compression techniques impact the network impedance determination. Finally, high voltage–direct current technologies are emerging candidates when electricity needs to be distributed over large distances. Consequently, modern interconnected power systems are becoming increasingly hybrid with both AC and DC subsystems. It is therefore of key importance to ensure the dynamic resilience of such hybrid systems in emergency situations through advanced control strategies. This topic is the central theme of the fifth contribution of this Special Issue.

Ultimately, the whole content of the Special Issue tackles research not only applied to the different voltage levels of power systems, but also focusing on different time scales (from steady-state disturbed operation to close-to-real-time stability matters), allowing one to have a good overview of the main research issues when dealing with the safe operation of modern power systems. Enjoy the reading!

François Vallée

Editor

Article

Peak Shaving through Battery Storage for Low-Voltage Enterprises with Peak Demand Pricing

Vasileios Papadopoulos [1,*], Jos Knockaert [1], Chris Develder [2] and Jan Desmet [1,*]

1 Lemcko, Department of Electromechanical, Systems and Metal Engineering, Ghent University, 8500 Kortrijk, Belgium; Jos.Knockaert@UGent.be

2 IDLab, Department of Information Technology, Ghent University—Imec, 9000 Gent, Belgium; Chris.Develder@UGent.be

* Correspondence: Vasileios.Papadopoulos@UGent.be (V.P.); Janj.Desmet@UGent.be (J.D.)

Received: 30 January 2020; Accepted: 1 March 2020; Published: 5 March 2020

Abstract: The renewable energy transition has introduced new electricity tariff structures. With the increased penetration of photovoltaic and wind power systems, users are being charged more for their peak demand. Consequently, peak shaving has gained attention in recent years. In this paper, we investigated the potential of peak shaving through battery storage. The analyzed system comprises a battery, a load and the grid but no renewable energy sources. The study is based on 40 load profiles of low-voltage users, located in Belgium, for the period 1 January 2014, 00:00–31 December 2016, 23:45, at 15 min resolution, with peak demand pricing. For each user, we studied the peak load reduction achievable by batteries of varying energy capacities (kWh), ranging from 0.1 to 10 times the mean power (kW). The results show that for 75% of the users, the peak reduction stays below 44% when the battery capacity is 10 times the mean power. Furthermore, for 75% of the users the battery remains idle for at least 80% of the time; consequently, the battery could possibly provide other services as well if the peak occurrence is sufficiently predictable. From an economic perspective, peak shaving looks interesting for capacity invoiced end users in Belgium, under the current battery capex and electricity prices (without Time-of-Use (ToU) dependency).

Keywords: peak shaving; battery storage; peak demand pricing; lithium-ion; tariff structure

1. Introduction

Over the past decade, most countries all over the world have taken action towards reducing their polluting emissions by investing in renewable energy sources. Among those sources, particularly, photovoltaic (PV) solar panels and wind power systems have seen a significant growth [1]. However, the increase of renewables goes hand in hand with technical challenges. The stochasticity of both PV and wind power systems causes the maintenance of grid stability to become more difficult [2,3].

A major stakeholder impacted by the renewable energy transition is the distribution network operator. While end users are becoming increasingly more independent from the grid, the revenue constraint for the grid operator still remains [4]. Under the current tariff structure, which is primarily based on the energy-volume component, a 'death spiral' phenomenon is imminent [4,5]. Nevertheless, the grid infrastructure costs are mainly dependent on the power capacity of the system. Yet, PV users have reportedly slightly lower peak power than non-PV users [6]. In other words, PV-users pay less than non-PV users even though both of them use the grid almost to the same extent [6]. To counteract such unfairness between different user groups and correctly attribute the costs to their origin, new tariff structures are being introduced that increase the weight factor for the peak demand component. This (peak demand pricing) will also apply for small user groups such as residential consumers who have been so far excluded from peak power measurements [7,8].

Given these increased peak power costs, peak demand reduction ('peak shaving') has gained much attention in recent years. Peak shaving is not a new concept; industrial users with high peak demand already have been using diesel and gas generators to reduce electricity costs for a long time. Still, those conventional generation methods are expected to be replaced by 'green' technologies, among which energy storage and in particular batteries are the primary candidate.

Battery storage systems have been deployed in the past to provide different types of services, such as (i) increasing the self-sufficiency of PV/wind power installations [9–11], (ii) providing ancillary services to the grid operator [12–14], (iii) peak shaving [15–17], (iv) back-up generators and UPS [18,19]. A common issue, arising particularly in (i), (ii) and (iii), is that due to the high cost of the storage system, battery storage investments are not yet economically feasible. However, we note that in the majority of those studies, the battery is deployed exclusively for one service. Therefore, to accelerate the return of investment, many suggest as a possible solution 'hybridizing' multiple services into a single application instead of providing each one separately [14,20,21]. Before studying how such a hybrid strategy can be applied, we should first identify the technical constraints of the services under consideration. In this paper, we focus specifically on peak shaving and present some insights that reflect its potential for hybridization. In the next paragraph, we review previous research works on peak shaving through battery storage.

In [15], the authors present a sizing methodology for defining the optimal energy and power capacity of battery storage systems used for peak shaving. An economic feasibility study was conducted for two different technologies, lead acid and vanadium redox flow (VRF). A control strategy was proposed, but it assumed that the load profile is perfectly predictable in advance. In [16], the researchers applied peak shaving for residential end users. One of the main conclusions was that the utilization of the lithium-ion battery stays very low, lower than 165 cycles per year. At such a low rate (here, the cycle lifetime is 3000 cycles) the system could be used for more than 20 years unless it exceeded its calendar lifetime. Finally, considering also its calendar lifetime, the battery would have to be replaced approximately after 10–15 years. Furthermore, the researchers suggested adding grid support services next to peak shaving in order to increase the utilization of the system. In [22], the researchers developed a model in Matlab/Simulink where a VRF battery is used to simultaneously provide frequency regulation and peak shaving. It was concluded that the battery storage system can successfully perform both services. However, the experiment was conducted only for a limited time period (30–140 s), thus, in essence, without affecting the battery state of charge (SoC) and as a consequence, it was not possible to evaluate the reliability of the control system under unfavorable conditions. In [23], a fuzzy control algorithm was developed for peak shaving in university buildings. The algorithm was tested and compared to two different peak shaving techniques, namely the fixed-threshold and adaptive-threshold controller. The results showed that the proposed algorithm was the best of all. Although the researchers conducted several case studies (with 8 different load profiles), they did not provide sufficient information about the load forecasting method. In [17], a control algorithm is proposed for peak shaving in low-voltage distribution networks based on day ahead aggregated load forecasts. The main novelty of that study is that the algorithm, considering also the inherent forecasting errors, relies solely on historic data; hence there is no need to intervene in real-time and readapt the dis-charging process of the battery. Results from a case study show that peak reduction is achieved for 97% of the time and that for 55% of the time, the peak reduction is at least 10%. In [18,19,24,25], peak shaving is addressed as a secondary application. Here, the primary service of the battery is to provide uninterruptible power supply (UPS) in data centers. The researchers argue that because of the significantly low probability of the peak occurrence (e.g., a Google data center exceeds 90% of its power capacity only for 1% of the time), it is possible to achieve peak reduction without impacting the reliability of the primary service. In [26], a battery sizing methodology and an optimal control algorithm is proposed for peak shaving in industrial and commercial customers. One of the main objectives was to determine an appropriate peak shaving threshold. Three case studies were carried out, each one considering a different daily load profile. The results showed that adapting the peak shaving threshold in real-time leads to higher peak reduction than keeping a fixed threshold based only on a historic data analysis. A drawback of the study might be that when calculating the battery utilization, it is assumed that the battery is equally utilized every weekday of the year, thus omitting possible idle periods on days with low power consumption. In [27], a peak shaving algorithm was proposed for microgrid applications. In contrast to conventional approaches considering only the load consumption, here, the peak threshold applies also for the PV generation. The battery capacity is equally reserved for both positive (injection to the grid) and negative (absorption from the grid) peaks by setting the SoC during normal operation at 50%. The algorithm was tested on a real-time microgrid, implemented in the lab. The researchers used predefined data (load/PV profiles) to carry out the experiment; however, they did suggest in future deploying predictive analytics to improve the reliability of the system.

In this paragraph, we explain three major contribution pillars of the present research work.

i. *Dataset:* First, an important conclusion to note, resulting from our literature review is that all previous studies refer to unique use cases. Moreover, in almost all previous studies, the data was very limited (max 2–3 months); thus, the seasonal periodicity was not present. To the best of our knowledge, the present study is the first to consider such large dataset: 40 load profiles (in the Supplementary Materials), each one with 3 full years of historic load power. Knowing the difficulties of finding qualitative data, we decided to make this dataset publicly available (The dataset is available as attachment to this manuscript. Or contact Vasileios.Papadopoulos@ugent.be) in order to stimulate further research on this topic.
ii. *Sizing methodology:* Secondly, aside from the extended datasets, another thing that has been missing from the existing literature on peak shaving, which has focused mainly on control strategies, is a concrete methodology of sizing the battery capacity. In the present paper, we demonstrate how to calculate the minimum battery capacity requirement by combining a power flow model with the dichotomy optimization algorithm.
iii. *Quantitative results:* Thirdly, in our attempt to strengthen the validity of our conclusions, we provide an overview of quantitative results from all 40 different use cases. We show both energetic assessments and economic results. The third contribution pillar can be summarized in answering the following:

    - How much peak demand reduction can a user achieve for a given battery energy capacity (kWh)?
    - What is the battery utilization, how much time during the year and how many cycles? Does peak shaving heavily impact the degradation of the battery? Can we hybridize peak shaving with other services?
    - Which performance metrics should we use and how can these be interpreted from an economic perspective? What are the profitability margins of battery storage for Belgium?

The rest of the paper is structured as follows. In Section 2, the data of the study are presented (Section 2.1). Then, we proceed with the methodology; the power flow model is explained (Section 2.2) and the dichotomy method is proposed as an optimization algorithm (Section 2.3). Section 2 closes with the definition of performance metrics (Section 2.4). Next, Section 3 shows the results of the simulation (Section 3.1) and explains how to interpret those from an economic perspective (Section 3.2). Finally, Section 4 summarizes the most important conclusions and makes suggestions for future research objectives.

2. Materials and Methods

2.1. Data

We received 40 load profiles from the Flemish distribution grid operator (Fluvius) Each profile is the active power (in kW) of an enterprise for the 3-year period between 1 January 2014, 00:00 and 31 December 2016, 23:45. All enterprises are low-voltage users with peak demand pricing and a connection capacity above 56 kVA and lower than 1 MVA. The data was logged through automatic measurement reading (AMR) devices with a time resolution of 15 min. The mean power of the users varied between 1.92 and 53.75 kW (Figure 1a). The peak-to-mean power ratio was between 1.5 and 40; however, for 90% of the users, the ratio is lower than 10 (Figure 1b).

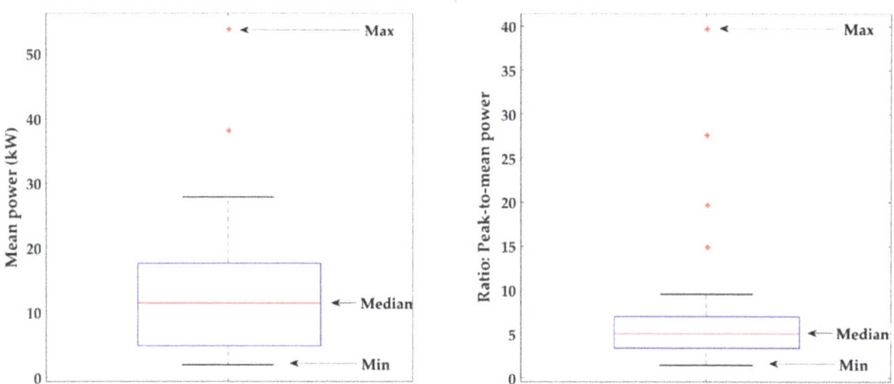

Figure 1. Boxplots, 40 load profiles–(**a**) Mean power (left), (**b**) Ratio: Peak-to-mean power (right).

2.2. Power Flow Model

Figure 2 shows the topology of our system. The battery is connected through a DC/AC inverter behind the meter of the user. The grid serves as the only power supply since there are no renewable energy sources. In general, for peak shaving, the energy storage system should have high energy efficiency as well as high power capacity (C rate) [28]. For these reasons, we selected a Lithium-ion battery to carry out our analysis (See Table 1).

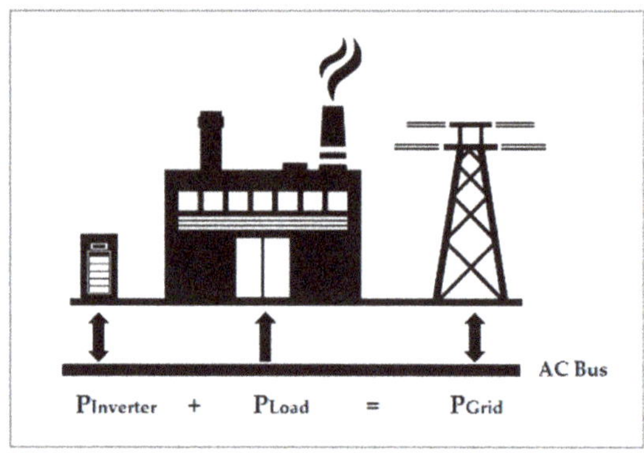

Figure 2. System topology.

Table 1. LFP Cell Characteristics, according to [29].

Characteristics	Specifications
Chemistry	$LiFePO_4$
Energy capacity	2.28 Ah (7.52 Wh)
Nominal voltage	3.3 V
Operating voltage	2.5 to 3.6 V
Operating temperature	−30 °C to + 60 °C
Cell weight	70 g

The simulation model, built in Matlab/Simulink is shown in Figure 4. Here, it is worth noting that a part of the present model used for peak shaving was based on the model described in [30]. Therefore, in this paper, we will only detail the new model components, which are blocks 1 and 5 (See Figure 4). For the remaining blocks 2, 3 and 4, we provide a generic description, but for more information, the reader is referred to [30], in particular its Section 2.3. For the development of the model, we relied heavily on a real test-setup—microgrid emulator (The microgrid emulator makes part of the laboratory infrastructure of EELab/Lemcko, an expertise center of Ghent university, specialized in Renewable Energy System applications. For more information, contact the first author (Vasileios.Papadopoulos@UGent.be)) comprising of: (i) a low-voltage grid (250 kVA power source), (ii) a 90 kVA DC/AC converter, (iii) a 20 kWh $LiFePO_4$ battery, (iv) a 30-kW programmable load. The behavior of each component and the interaction between them was studied analytically and converted into simulation models using information from test measurements, scientific papers and commercial datasheets.

To begin with, the model has three variables: (i) the time resolution of the load profile, (ii) the battery capacity (kWh) and (iii) its C rate. Furthermore, it receives two data inputs: (i) the load profile and (ii) a power threshold. The load profile is simply a time series of the active power in kW at 15 min resolution. The power threshold is a constant specifying the 'desired' maximum power. This value must be lower than the peak power but also higher than the mean power. Given the time step (resolution) and the 3-year period, in total, there are 105,216 simulation steps (1096 days × 96 quarters/day). At each step, the model reads the load power of that moment and the current State-of-Charge (SoC). Then, it undergoes three sequential processes (1, 2 and 3) to calculate the battery power P_{bat} (inverter's DC side), the inverter power P_{inv} (inverter's AC side) and the power of the grid P_{grid}. Next, after updating the State-of-Charge (SoC) of the battery, it proceeds to the next simulation step and hence, the simulation progresses. Figure 3 shows the DC/AC conversion efficiency of the inverter in charging mode. Additionally, all the equations that were used to calculate the inverter power P_{inv} and battery power P_{bat} in charging and discharging mode.

$$P_{bat} = f(x) \cdot P_{inv} \tag{1}$$

$$\frac{P_{bat}}{P_{nom}} = f(x) \cdot \frac{P_{inv}}{P_{nom}} \tag{2}$$

$$\frac{P_{bat}}{P_{nom}} = f(x) \cdot x = g(x) \tag{3}$$

$$\frac{P_{inv}}{P_{nom}} = g^{-1}(\frac{P_{bat}}{P_{nom}}) \tag{4}$$

$$P_{bat} = \frac{P_{inv}}{f(x)} \tag{5}$$

$$\frac{P_{bat}}{P_{nom}} = \frac{P_{inv}}{P_{nom}} \cdot \frac{1}{f(x)} \tag{6}$$

$$\frac{P_{bat}}{P_{nom}} = \frac{x}{f(x)} = h(x) \tag{7}$$

$$\frac{P_{inv}}{P_{nom}} = h^{-1}\left(\frac{P_{bat}}{P_{nom}}\right) \quad (8)$$

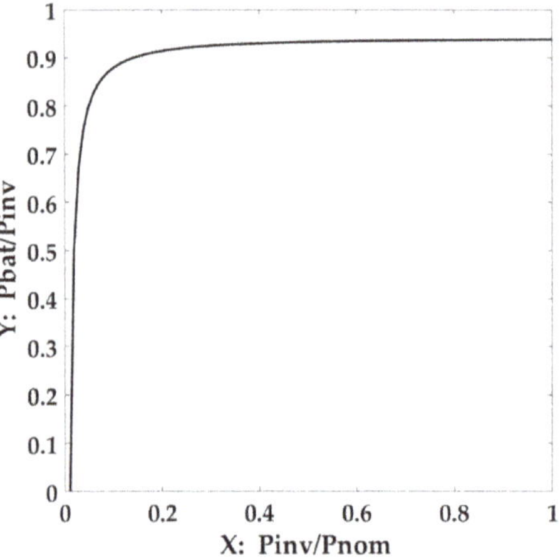

Figure 3. DC/AC efficiency, Y = f(x).

With respect to the sequential processes, process 1 performs the power conversion from AC to DC compensating for the efficiency losses (AC to DC). Process 2 applies two saturation constraints to the battery power: one for the given C rate and one for the given time resolution. Finally, process 3 performs the reverse conversion from DC to AC considering the inverse (DC to AC) efficiency losses. In the following paragraph, we describe with more detail those processes.

Process 1—AC/DC power conversion (Figure 4, block 1): Initially, we set the inverter power equal to the difference $P_{Threshold} - P_{load}$. In case of a power surplus (positive difference), the inverter is in charging mode to restore the battery's energy capacity, otherwise, in case of a power deficit (negative difference), the inverter is in discharging mode to shave the peak. After setting the inverter power, next, we calculated the battery power compensating for the efficiency losses. In charging mode, the battery power is always lower than the inverter power (See Equation (1)) and vice versa in discharging mode the battery power is always higher than the inverter power (See Equation (5)).

Process 2—Power saturation constraints (Figure 4, block 2, 3, 4): Here, we impose two constraints to the battery power. First (block 2), the battery power can never exceed its power capacity as specified by its C rate limit and the SoC level. For this battery technology, the recommended C rate is 1. How we calculate exactly the power from the C rate limit, has been explained in [30], Section 2.3. (As an approximation, we can state that the power capacity is equal to the battery's nominal voltage times the C rate, times its energy capacity in Ah: $P_{bat\ max} = U_{nom} \cdot C_{rate} \cdot C_{Ah} \cdot$) Second (block 3), we must take into account also the time resolution of our data (15 min). This constraint comes into effect when the SoC level is very close either to its upper or lower limit (90% and 10% respectively) (10–90% is the recommended by the manufacturer SoC range to maximize the lifetime of the battery). Since our simulation is executed in discrete steps of 15 min, we need to consider how much energy is left inside the battery and saturate its power accordingly (see [30], Section 2.3). Afterwards, at the output of the second constraint, the battery power was finally defined and hence the SoC can be updated (block 4).

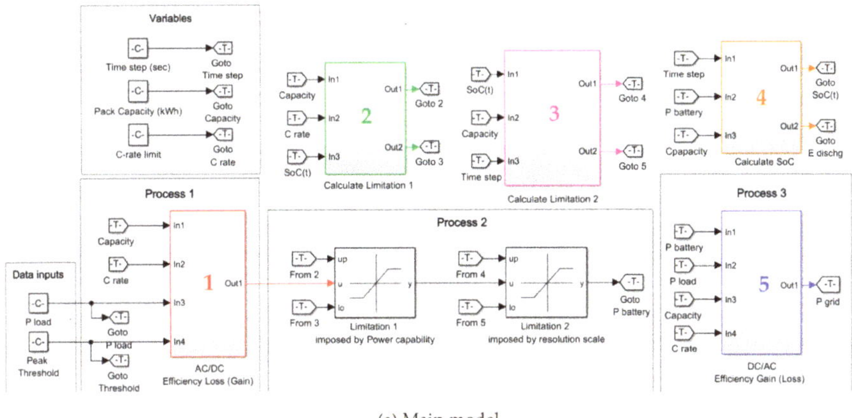

(a) Main model

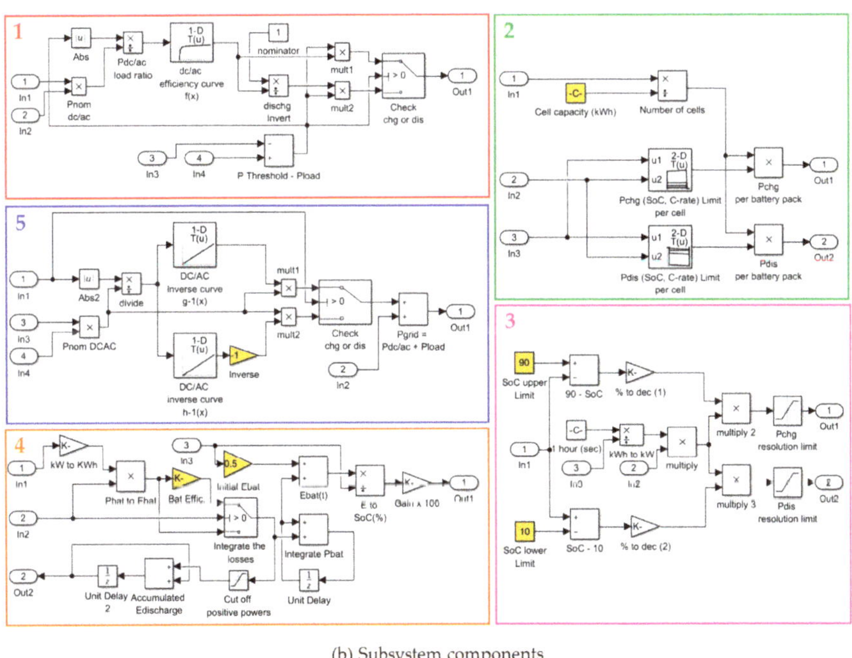

(b) Subsystem components

Figure 4. Power flow model for peak shaving designed in Matlab/Simulink.

Process 3—DC/AC power conversion (Figure 4, block 5): Knowing the final value of the battery power, it is then possible to calculate the final value of the inverter power. At this point, the DC/AC efficiency function f(x) needs to be inverted. In charging mode, we make use of Equation (4) (function g^{-1}) and in discharging mode Equation (8) (function h^{-1}). As a result, we finally know both the load power P_{load} and the inverter power P_{inv}. Therefore, we can also calculate the power of the grid P_{grid} ($P_{grid} = P_{load} + P_{inv}$) and proceed to the next simulation step.

2.3. Dichotomy Method

As already mentioned in Section 2.2, the Simulink model receives both the battery capacity (as variable) and a peak threshold (as data input). To find out whether or not that threshold will be met, all we have to do is run the simulation and check the maximum load power $\text{Max}(P_{\text{load}})$. On the one hand, if the threshold is too low, the system will be unreliable ($\text{Max}(P_{\text{load}}) > P_{\text{Threshold}}$) due to insufficient battery capacity, whereas, on the other hand, if the threshold is too high ($\text{Max}(P_{\text{load}}) \leq P_{\text{Threshold}}$) the system will be reliable but the battery is overdimensioned. Consequently, for each load profile and a given battery capacity, there is only one threshold that minimizes the load power (See Figure 5). To find the solution for our optimization problem we deployed the 'dichotomy method'. In the next paragraph, follows a short description of the algorithm.

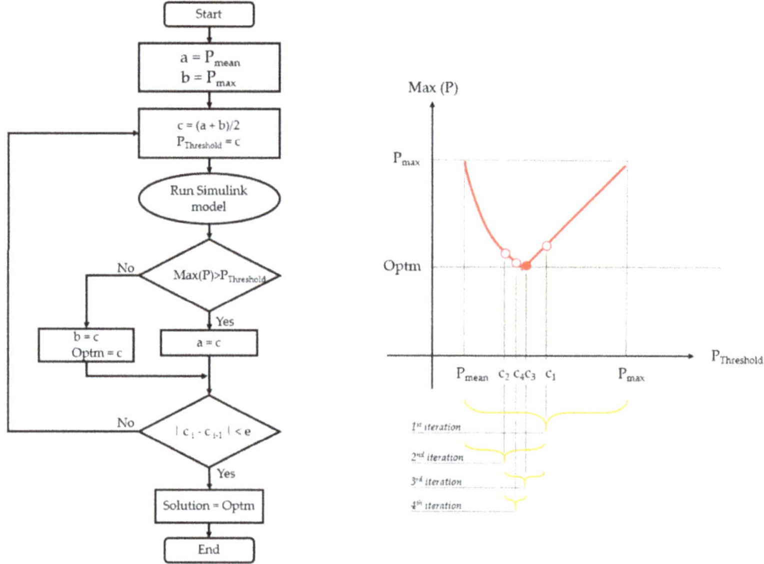

Figure 5. Flow chart—Dichotomy method: (**a**) Pseudocode (left), (**b**) Midpoint evolution (right).

Dichotomy method (Figure 5):

1. Initialize the lower and upper threshold limit at a = P_{mean} and b = P_{max}, respectively.
2. Enter dichotomy loop: Calculate the midpoint at c = (a + b)/2 and set the peak threshold equal to that value.
3. Run the Simulink model.
4. Check the maximum load power. If the load power exceeds the threshold update the lower limit at a = c. Otherwise, update the upper limit at b = c and store that value as the current solution.
5. Check convergence criterion. If the distance between the current and previous midpoint is lower than a constant, exit the loop, otherwise, go to step 2 and recalculate the new midpoint.

2.4. Definition of Performance Metrics

Before continuing with the presentation of the simulation results, first, we need to give the definitions of our performance metrics, based on which we evaluated the peak shaving potential of the users. In our approach, we would rather associate the word 'potential' explicitly to energetic assessments. The extent to which these can be translated into economic terms (e.g., revenues, expenses, ROIs) depends certainly on the tariff structure under consideration as well as the cost for the battery storage system. Although, as shown in Section 3, we do provide some insights specifically for Belgium, preferably, each reader ought to make his own reflections.

Peak reduction (%): It is the percentual difference between the initial peak power and the final peak power after peak shaving:

$$A_{peak\ red} = \frac{P_{max\ i} - P_{max\ f}}{P_{max\ i}} \cdot 100 \tag{9}$$

where $A_{peak\ red}$ is the peak reduction, $P_{max\ i}$ is the initial peak power, $P_{max\ f}$ is the final peak power after peak shaving.

Peak reduction-to-capacity: It is the difference between the initial peak power and the final peak power after peak shaving divided by the battery capacity. This metric can serve us as a rough estimation of the profitability of the installation if we can express the revenue and costs linearly proportional to the peak reduction and battery capacity respectively.

$$R_{peak\ red-to-cap} = \frac{P_{max\ i} - P_{max\ f}}{C_{bat}} \tag{10}$$

where $R_{peak\ red-to-cap}$ is the ratio peak reduction-to-capacity, $P_{max\ i}$ is the initial peak power, $P_{max\ f}$ is the final peak power after peak shaving, C_{bat} is the battery capacity.

SoC active time (%): It is the average percentage of time per year that the battery is deployed for peak shaving. This metric can be very useful, especially when our intention is to combine peak shaving with other services (e.g., increasing the self-sufficiency of PV, ancillary services, Time-of-Use (ToU) prices).

$$\begin{aligned} SoC_{act\ time} &= \sum_{i=1}^{1096 \cdot 96} i \cdot \frac{100}{1096 \times 96} \\ i &= \begin{cases} 1, & |P_{bat}| > 0 \\ 0, & P_{bat} = 0 \end{cases} \end{aligned} \tag{11}$$

where $SoC_{act\ time}$ is the SoC active time, P_{bat} is the battery power, i is the quarter index of the simulation, 1096 × 96 is the total number of quarters within the 3 years period (1st January 2014–31st December 2016).

Battery utilization (cycles/year): It is the average total energy discharged by the battery within a year divided by the battery capacity. This metric can be used to assess how fast the battery reaches the end of its lifetime. Particularly for peak shaving applications, it is desirable that the battery be utilized as low as possible since our cost savings are exclusively dependent on the power component (cost in function of kW). Conversely, when the aim is to increase the self-sufficiency of the installation (PV or wind), the battery utilization should be as high as possible, since our cost savings are mainly dependent on the energy component (cost in function of kWh).

$$U_{bat} = \frac{E_{dis\ tot}}{C_{bat} \cdot 3} \tag{12}$$

where U_{bat} is the battery utilization, $E_{dis\ tot}$ is the total discharged energy within the 3 years period, C_{bat} is the battery capacity.

Consumption increase (%): It is the percentage of energy consumption increase due to efficiency losses of the battery storage system. In addition to the initial capital expenditures for the battery, the additional energy consumption should be taken into account as operating cost.

$$A_{incr} = \frac{E_{load\ f} - E_{load\ i}}{E_{load\ i}} \cdot 100 \tag{13}$$

where A_{incr} is the consumption increase, $E_{load\ f}$ and $E_{load\ i}$ is the total energy consumed within the 3-year period after and before peak shaving, respectively.

3. Results

As mentioned in Section 2, the power flow model receives three variables: battery capacity, C rate, and time step. For each load profile in our dataset, we carried out multiple simulations by varying only the battery capacity, whereas both the time step and the C rate were set at constant values. The peak threshold was calculated using the dichotomy method after defining the battery capacity.

The time step was set at 15 min which is the time resolution of the dataset. The C rate was set at 1 C; higher values are not recommended for the chosen battery technology because this would negatively impact its lifetime. Furthermore, based on our experience, for most applications, 1 C is sufficiently high to meet a given peak threshold. In general, the extent to which we can reduce the peak power depends on the battery's energy capacity rather than its power capacity. Nevertheless, we do suggest for future research to investigate the impact of the C rate as well, but in this study, it will not be addressed. Regarding the battery capacity, since we deal with several users, in order to maintain a common reference of comparison between the users, we normalized the battery capacity by dividing it by the mean power of the user. Finally, the ratio battery capacity-to-mean power was varied within 0.1–10.

3.1. Energetic Assessments

The simulation results are presented in Figures 6 and 7. Knowing that our dataset consists of 40 users, it would be ineffective to illustrate 40 individual plots into the same figure. Instead, we selected five quantile elements at which the cumulative probability becomes 5%, 25%, 50%, 75% and 95%. This gives us a better view of the statistical distribution of each performance metric.

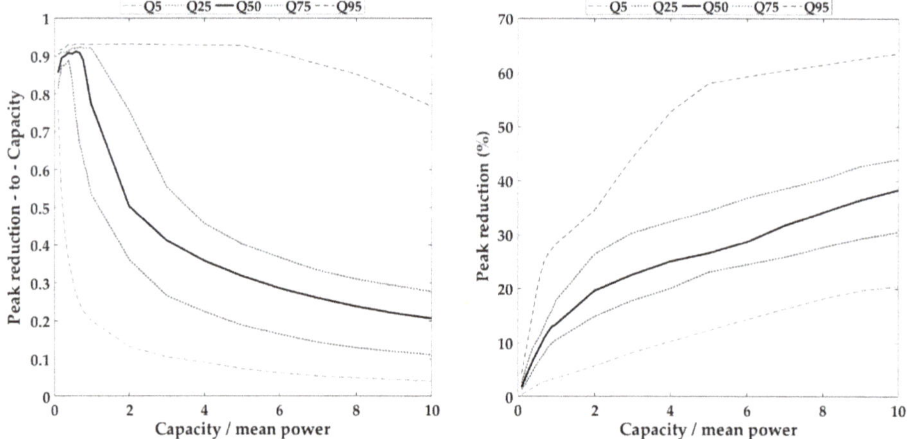

Figure 6. Simulation results: (**a**) Peak reduction-to-capacity (left), (**b**) peak reduction (right).

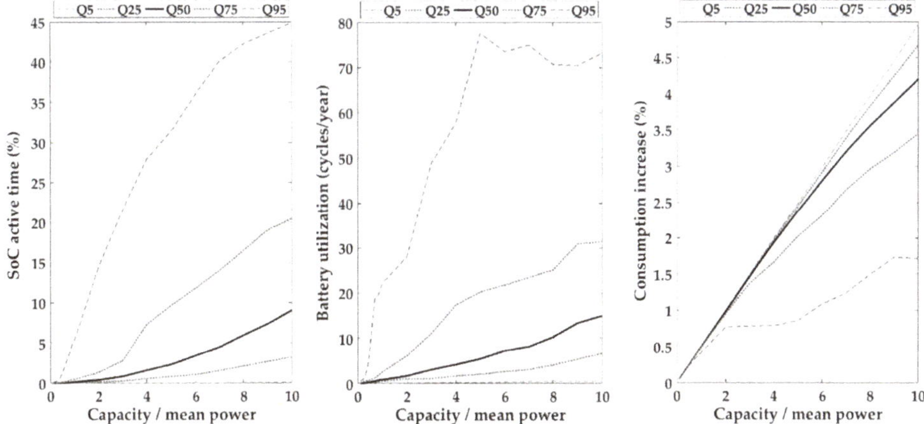

Figure 7. Simulation results: (**a**) SoC active time (left), (**b**) Battery utilization (middle), (**c**) Consumption increase (right).

From both Figure 6a,b, it can be concluded that the peak reduction increase decreases with the battery capacity (second derivative of the function in Figure 6b is negative) or in other words: as the battery capacity increases, peak shaving becomes more difficult. For a battery capacity 2 times the mean power (e.g., a user with 30 kW mean power installs a 60 kWh battery) seventy percent of the users between Q5 and Q75 achieve peak reduction in the range 0.26–1.5 times their mean power (Figure 6a). The same group of users achieves peak reduction up to 6–27% of their peak power (Figure 6b). For a battery capacity of 10 times the mean power (e.g., a user with 30 kW mean power installs a 300 kWh battery) the peak reduction for that group (Q5–Q75) varies within 0.4–2.8 times their mean power (Figure 6a) and 20–44% of their peak power (Figure 6b).

Regarding the SoC active time (Figure 7a), it increases with the battery capacity. The reason is that as the battery capacity increases, the peak threshold is reduced and consequently, the battery is used more frequently. An important conclusion to note is that, for most users, the SoC active time remains very low, even for large battery capacities. Seventy percent of the users between Q5 and Q75 with a battery capacity 10 times the mean power deploy their battery in the range of 0–20%, or in other words the battery stays idle for at least 80% of the time during the year. This fact in itself opens up new research opportunities.

If peak shaving does occur rarely, then we could possibly hybridize our energy management system including other services as well (e.g., ancillary services, increasing the self-sufficiency of renewable energy installations). Figure 7b provide another indication that the battery is underutilized, here, however in terms of lifetime expectancy. Over the entire battery capacity dimension, for ninety-five percent of the users (Q0–Q95), the battery does not deliver more than 80 cycles per year. This number is considerably lower compared to the cycle lifetime of nowadays' state-of-the-art Lithium-ion technologies (above 5000 cycles) [28]. At such low utilization rates, the battery can endure several years of use (more than a decade). Finally, it will be due to another reason why the battery was decommissioned such as a maintenance issue or simply because the battery has reached the end of its calendar lifetime. (The capacity fade effect of Lithium-ion batteries is both time-dependent (calendar lifetime) and cycle-number dependent (cycle lifetime). Regardless of its utilization, after a certain time period the battery loses a part of its initial capacity. Usually, the End of Life (EoL) of a battery is defined when its initial capacity is reduced by 20%, in many critical applications (e.g., EVs) this is the time when the battery needs to be either decommissioned or repurposed for another application.)

The consumption increase is shown in Figure 7c. It is worth noting once more that the battery technology in the present study exhibits a very high energy efficiency. Undoubtedly, if other technologies were used (e.g., lead acid, flow batteries), the consumption increase would be higher. As can be seen from the figure, obviously, the higher the battery capacity, the higher the absolute energy losses. One reason why this happens is due to the increase of the battery utilization (see Figure 7b) and another reason is because both the battery and the dc/ac converter become bigger in size. Consider, for instance, a user with 30 kW mean power and a battery capacity of 300 kWh (capacity-to-mean power is 10). Only the converter losses to (dis)charge the battery at 1 C are approximately 15 kW (at 95% dc/ac efficiency). If the battery capacity was 30 kWh (capacity-to-mean power is 1), those losses would be significantly lower (1.5 kW).

3.2. Economic Evaluations

Let us now consider a case study of how to interpret those results from an economic perspective. Table 2 lists the parameters used for our economic analysis:

- The electricity price is an average for Belgium energy invoices in the considered capacity connection range. The electricity price lies in the range of 0.2–0.25 €/kWh [31]. Here, it must be noted that our analysis is exclusively applicable for end users with fixed electricity prices during the year; there is no Time-of-Use (ToU) dependency (e.g., day/night tariff, dynamic pricing schemes). (In case of ToU dependency, the control strategy of the battery is different. Peak demand pricing coexists with ToU pricing and therefore, we need to solve the economic optimization problem first.)
- Regarding the peak shaving compensation, for the DSO in Belgium, peak demand is defined as the highest 15 min load power measured by the user's AMR meter over the last 12 months. The compensation ranges approximately within 87.6–131 €/kW per year depending on the geographical location. By dividing by the total number of hours per year (8760 h), this equals 0.01–0.015 €/kW/h. (These values have been defined using a cost simulation tool from the distribution grid operator. The values apply exclusively to those users connected to the low-voltage grid with peak demand pricing.)
- With respect to the battery storage system, we consider capital expenditures at 500 €/kWh (per kWh of energy capacity). The consumption increase can be approximated as linear function of the battery capacity (See Figure 7c) at 0.4%/capacity-to-mean power. The battery cycle lifetime is estimated at 5000 cycles (at 80% EoL) considering normal operating conditions: (i) Ambient temperature 25 °C, (ii) SoC within 10–90%, (iii) (dis)charge current at 1C. Given that our battery utilization is very low (80 cycles/year worst case), we will consider only calendar aging at 2% capacity loss per year. (To calculate the battery's cycle lifetime and calendar aging, under those conditions (25 °C, 10–90% SoC, 1C) we received information from the manufacturer. For those interested in analytic methods to calculate the battery cycle lifetime and calendar aging, we refer to noteworthy research works [32,33].) The payback period of our investment is 10 years and we do not consider any option to resell the battery; after this period the battery is recycled.

Table 2. Peak Shaving—Parameters for Economic Feasibility.

Parameters	Values
Payback period	10 Years
Peak shaving compensation	0.01–0.015 €/kW/h
Battery capex	500 €/kWh
Consumption increase rate	0.4%/capacity-to-mean power
Electricity price	0.20–0.25 €/kWh
Battery capacity fade	2% per year

In order for the system to be profitable, the total peak shaving compensation has to be higher than the total cost (incl. battery and losses) over the payback period; this condition is expressed in Equation (14). Next, as shown in Equation (15), the peak reduction-to-capacity ratio can be expressed in function of all economic parameters. Finally, by replacing with the values of Table 2, it can be concluded that the ratio needs to be higher than 0.43–0.67 (Equation (16)).

$$\text{Rev}\cdot 8760\cdot \text{ROI}\cdot \Delta P_{peak} > \text{Cap}\cdot(\text{Cost}_{bat} + \text{Rate}_{cons\ incr}\cdot P_{elect}\cdot 8760\cdot \text{ROI}) \quad (14)$$

$$\frac{\Delta P_{peak}}{\text{Cap}} > \frac{\text{Cost}_{bat} + \text{Rate}_{cons\ incr}\cdot P_{elect}\cdot 8760\cdot \text{ROI}}{\text{Rev}\cdot 8760\cdot \text{ROI}} \quad (15)$$

$$\frac{\Delta P_{peak}}{\text{Cap}} > 0.43 - 0.67 \quad (16)$$

where ΔP_{peak} is the peak reduction, ROI is the return of investment (payback period), Rev is the peak compensation (revenue), Cap is the battery capacity, Cost_{bat} is the battery capex, $\text{Rate}_{cons\ incr}$ is the rate of consumption increase and P_{elect} is the electricity price.

Over the 10-year period, the total capacity loss of the battery will be 20%. Consequently, to ensure that the peak threshold will always be met, we have to oversize the battery capacity. Finally, the results of the economic feasibility study are illustrated in Figure 8. Figure 8 can be made easily from Figure 6a (see Section 3.1) by adding a 20% margin to the minimum battery capacity requirement. The color at each point [x,y] represents the total number of users whose peak reduction-to-capacity exceeds the y value (similarly to the quantile plots of Figure 6a). The yellow and green dashed lines represent the profitability thresholds 0.43 and 0.67, respectively (see Equation (16)). As can be seen, there are several positive use cases; of course the number of positive cases depends on the battery size. To give an example, when the ratio capacity/mean power equals 2, there are 15–20 users exceeding the value 0.43 (lower profitability threshold), whereas when the ratio capacity/mean power becomes 10, there are only 0–5 users exceeding that value (0.43). With that being said, we do now have an estimation of the profitability margins for the Belgian use cases.

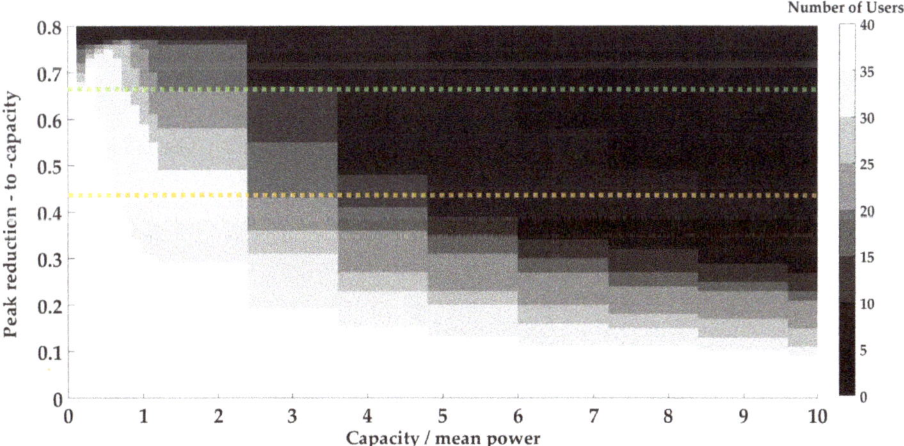

Figure 8. Peak shaving—results of economic feasibility study. At each point [x, y], the color represents the total number of users whose peak reduction-to-capacity exceeds the y value. The yellow and green dashed lines represent the profitability thresholds 0.43 and 0.67, respectively (see Equation (16)).

Needless to say that our estimation is strongly influenced by the considered parameter values (Table 2). Even without changing neither the electricity price nor the peak shaving compensation, simply by varying the payback period and/or the battery capex we would get different results. Here, it is worth noting that the battery capex at 500 €/kWh is very realistic for the time being and it is expected to decline further in the coming years [34]. (To define the battery capex we consulted manufacturers and received offers.) As a general conclusion, we can note that given the current electricity prices (fixed, no ToU dependency) and capital expenditures, particularly for Belgium, peak shaving through battery storage seems to be interesting from an economic perspective for several low-voltage enterprises.

4. Discussion and Conclusions

To summarize briefly what has been done, a model was developed in Matlab/Simulink for peak shaving. The dichotomy method was proposed as an optimization algorithm to find the minimum threshold above which we are 100% certain that the peak will never be exceeded. The model was tested for 40 low-voltage users with peak demand charge derived from the Belgian grid operator. We introduced five performance metrics to evaluate the simulation results. Furthermore, we gave an example how to interpret the results from economic perspective and explored the profitability of the application in Belgium. Below is a summary of the most important conclusions resulting from our analysis:

- For a battery capacity 2 times the mean power, the peak reduction of the group of users Q5–Q75 varies between 6% and 27%, whereas for a battery capacity 10 times the mean power, the peak reduction ranges between 20% and 44%.
- The SoC active time is limited for almost all cases. Even with an over-dimensioned battery (capacity-to-mean power is 10), for seventy-five percent of the users (Q0–Q75), the battery remains idle for at least 80% for the time. Consequently, peak shaving could possibly be hybridized with other services (e.g., increasing PV self-sufficiency, ancillary services) in order to accelerate the return of investment of the battery storage system. (By adding more revenue streams (stacked services) the payback period of the investment can be reduced.)
- The battery utilization is very low, up to 80 cycles per year in worst case. This number is significantly lower compared to the cycle lifetime of nowadays' lithium-ion batteries.
- The consumption increase gets higher with the battery capacity. It lies in the range 0% to 5% and does not substantially impact the operating cost of the system.
- From an economic perspective, peak shaving seems to be interesting for several low-voltage users in Belgium under the current capex and fixed electricity prices (no ToU dependency).

One of our main conclusions is that the battery utilization (SoC active time and number of cycles) is very low for almost all users. Consequently, there seems to be enough potential to let our battery provide additional services during those inactive periods in order to accelerate the payback period of our investment. Which services can be combined and how efficiently this can be done is certainly a topic to be addressed by future research works.

As an initial step, we suggest studying the predictability of the load profile. In our study, we consider the battery to be available for peak shaving 100% of the time; therefore, there is no need to know in advance when the peak occurs. However, in hybrid applications, time must be allocated appropriately and as a result load prediction plays an important role. To better explain this argument, let us consider two different load profiles derived from our dataset, user A and B (Figures 9 and 10 respectively). Although the battery utilization is in both cases very low (peak occurs rarely), user A is by far more predictable than the user B. For user A, the peak occurrence is dependent on the day, the time of use and the temperature, whereas for user B, there seem to be no clear explanatory variables. Consequently, user B cannot know how to allocate his inactive time to other services; hence, the battery remains underutilized solely reserved for peak shaving. Closing this paragraph, we note that, so far, most research works on battery storage have addressed only single applications. In our view, the concept of hybridization will gain more attention in the coming years as users gradually acquire more incentives to interact with the grid.

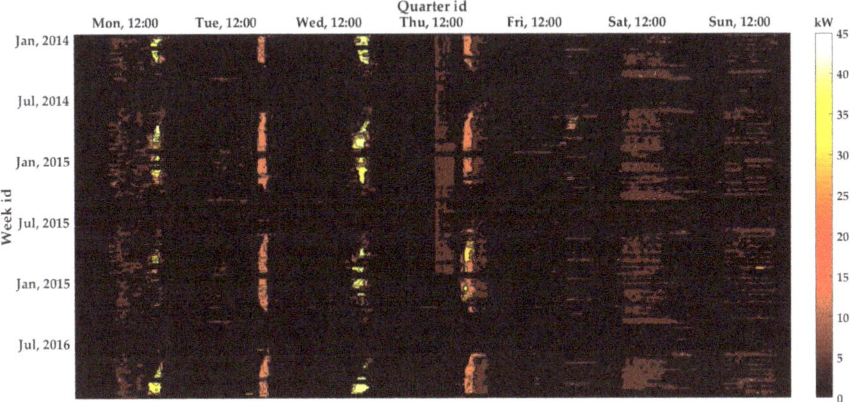

Figure 9. Thermal image—predictable load profile.

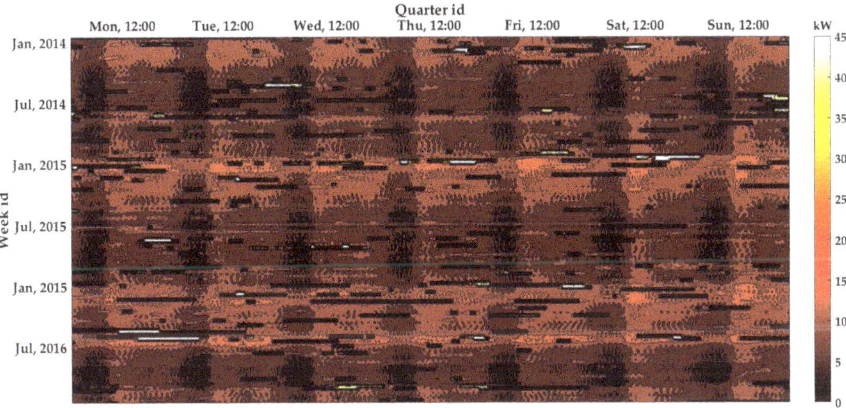

Figure 10. Thermal image—unpredictable load profile.

Supplementary Materials: The following are available online at http://www.mdpi.com/1996-1073/13/5/1183/s1, Dataset S1: "40 LV consumers with peak demand pricing in Belgium, From 1-1-2014 to 31-12-2016.zip".

Author Contributions: Conceptualization, V.P. and J.D.; methodology, V.P. and J.D.; software, V.P.; validation, V.P.; formal analysis, V.P.; investigation, V.P. and J.D.; resources, J.D.; data curation, C.D.; writing—original draft preparation, V.P.; writing—review and editing, J.D., J.K. and C.D.; visualization, V.P.; supervision, J.D., J.K. and C.D.; project administration, V.P. and J.D.; funding acquisition, J.D. All authors have read and agreed to the published version of the manuscript.

Funding: The research received no external funding.

Acknowledgments: We would like to thank the Flemish distribution system operator for providing the dataset for experimentation. Normally, such large datasets are very difficult to find within the research community.

Conflicts of Interest: The authors declare no conflict of interest.

References

1. Arantegui, R.L.; Jager-Waldau, A. Photovoltaics and wind status in the European Union after the Paris Agreement. *Renew. Sustain. Energy Rev.* **2018**, *81*, 2460–2471. [CrossRef]
2. Petinrin, J.O.; Shaaban, M. Impact of renewable generation on voltage control in distribution systems. *Renew. Sustain. Energy Rev.* **2016**, *65*, 770–783. [CrossRef]

3. Antonelli, M.; Desideri, U.; Franco, A. Effects of large scale penetration of renewables: The Italian case in the years 2008–2015. *Renew. Sustain. Energy Rev.* **2018**, *81*, 3090–3100. [CrossRef]
4. Simshauser, P. Distribution network prices and solar PV: Resolving rate instability and wealth transfers through demand tariffs. *Energy Econ.* **2016**, *54*, 108–122. [CrossRef]
5. Lummi, K.; Rautiainen, A.; Jarventausta, P.; Heine, P.; Lehtinen, J.; Hyvarinen, M. Cost-causation based approach in forming power-based distribution network tariff for small customers. In Proceedings of the 2016 13th International Conference on the European Energy Market (EEM), Porto, Portugal, 6–9 June 2016; pp. 1–5.
6. Strielkowski, W.; Streimikiene, D.; Bilan, Y. Network charging and residential tariffs: A case of household photovoltaics in the United Kingdom. *Renew. Sustain. Energy Rev.* **2017**, *77*, 461–473. [CrossRef]
7. VREG. *Consultatiedocument Tariefstructuur Periodieke Distributienettarieven Elektriciteit Voor Klanten Met een Kleinverbruiksmeetinrichting*; VREG: Flanders, Belgium, 2019.
8. VREG. *Consultatiedocument Tariefstructuur Periodieke Distributienettarieven Elektriciteit Voor Klanten Met een Grootverbruiksmeetinrichting*; VREG: Flanders, Belgium, 2019.
9. Nyholm, E.; Goop, J.; Odenberger, M.; Johnsson, F. Solar photovoltaic-battery systems in Swedish households—Self-consumption and self-sufficiency. *Appl. Energy* **2016**, *183*, 148–159. [CrossRef]
10. Silva, G.E.; Hendrick, P. Lead-acid batteries coupled with photovoltaics for increased electricity self-sufficiency in households. *Appl. Energy* **2016**, *178*, 856–867. [CrossRef]
11. Bertsch, V.; Geldermann, J.; Luhn, T. What drives the profitability of household PV investments, self-consumption and self-sufficiency? *Appl. Energy* **2017**, *204*, 1–15. [CrossRef]
12. Huvilinna, J. *Value of Battery Energy Storage at Ancillary Service Markets*; Aalto University: Espoo, Finland, 2015.
13. Koller, M.; Borsche, T.; Ulbig, A.; Andersson, G. Review of grid applications with the Zurich 1 MW battery energy storage system. *Electr. Power Syst. Res.* **2015**, *120*, 128–135. [CrossRef]
14. Staffell, I.; Rustomji, M. Maximising the value of electricity storage. *J. Energy Storage* **2016**, *8*, 212–225. [CrossRef]
15. Oudalov, A.; Cherkaoui, R.; Beguin, A. Sizing and optimal operation of battery energy storage system for peak shaving application. In Proceedings of the 2007 IEEE Lausanne Power Tech, Lausanne, Switzerland, 1–5 July 2007. [CrossRef]
16. Leadbetter, J.; Swan, L. Battery storage system for residential electricity peak demand shaving. *Energy Build.* **2012**, *55*, 685–692. [CrossRef]
17. Rowe, M.; Yunusov, T.; Haben, S.; Singleton, C.; Holderbaum, W.; Potter, B. A Peak Reduction Scheduling Algorithm for Storage Devices on the Low Voltage Network. In *IEEE Transactions on Smart Grid*; IEEE: Piscataway, NJ, USA, 2014; Volume 5, pp. 2115–2124. [CrossRef]
18. Govindan, S.; Wang, D.; Sivasubramaniam, A.; Urgaonkar, B. *Leveraging Stored Energy for Handling Power Emergencies in Aggressively Provisioned Datacenters*; ACM Sigplan Notices: New York, NY, USA, 2012; pp. 75–86.
19. Kontorinis, V.; Zhang, L.E.; Aksanli, B.; Sampson, J.; Homayoun, H.; Pettis, E.; Tullsen, D.M.; Rosing, T.S. Managing Distributed UPS Energy for Effective Power Capping in Data Centers. In Proceedings of the 2012 39th Annual International Symposium on Computer Architecture (ISCA), Portland, OR, USA, 9–13 June 2012; pp. 488–499.
20. Fitzgerald, G.; Mandel, J.; Morris, J.; Touati, H. *The Economics of Battery Energy Storage: How Mutli-Use, Customer-Sited Batteries Deliver the Most Services and Value to Customers and the Grid*; Rocky Mountain Institute: New York, NY, USA, 2015.
21. Smartest Energy. *Making the Commercial Case for Battery Storage*; Smartest Energy: London, UK, 2016.
22. Lucas, A.; Chondrogiannis, S. Smart grid energy storage controller for frequency regulation and peak shaving, using a vanadium redox flow battery. *Int. J. Electr. Power* **2016**, *80*, 26–36. [CrossRef]
23. Chua, K.H.; Lim, Y.S.; Morris, S. A novel fuzzy control algorithm for reducing the peak demands using energy storage system. *Energy* **2017**, *122*, 265–273. [CrossRef]
24. Aksanli, B.; Rosing, T.; Pettis, E. Distributed Battery Control for Peak Power Shaving in Datacenters. In Proceedings of the 2013 International Green Computing Conference Proceedings, Arlington, VA, USA, 27–29 June 2013.
25. Palasamudram, D.S.; Sitaraman, R.K.; Urgaonkar, B.; Urgaonkar, R. Using batteries to reduce the power costs of internet-scale distributed networks. In Proceedings of the Third ACM Symposium on Cloud Computing, San Jose, CA, USA, 14–17 October 2012; pp. 1–14.

26. Chua, K.H.; Lim, Y.S.; Morris, S. Energy storage system for peak shaving. *Int. J. Energy Sect. Manag.* **2016**, *10*, 3–18. [CrossRef]
27. Garcia-Plaza, M.; Carrasco, J.E.G.; Alonso-Martinez, J.; Asensio, A.P. Peak shaving algorithm with dynamic minimum voltage tracking for battery storage systems in microgrid applications. *J. Energy Storage* **2018**, *20*, 41–48. [CrossRef]
28. Guney, M.S.; Tepe, Y. Classification and assessment of energy storage systems. *Renew. Sustain. Energy Rev.* **2017**, *75*, 1187–1197. [CrossRef]
29. Ansean, D.; Gonzalez, M.; Viera, J.C.; Garcia, V.M.; Blanco, C.; Valledor, M. Fast charging technique for high power lithium iron phosphate batteries: A cycle life analysis. *J. Power Sources* **2013**, *239*, 9–15. [CrossRef]
30. Papadopoulos, V.; Knockaert, J.; Develder, C.; Desmet, J. Investigating the need for real time measurements in industrial wind power systems combined with battery storage. *Appl. Energy* **2019**, *247*, 559–571. [CrossRef]
31. VREG. Evolutie Energieprijzen en Distributienettarieven. Available online: https://www.vreg.be/nl/evolutie-energieprijzen-en-distributienettarieven (accessed on 10 October 2019).
32. Liu, K.L.; Hu, X.S.; Yang, Z.L.; Xie, Y.; Feng, S.Z. Lithium-ion battery charging management considering economic costs of electrical energy loss and battery degradation. *Energy Convers. Manag.* **2019**, *195*, 167–179. [CrossRef]
33. Liu, K.; Li, Y.; Hu, X.; Lucu, M.; Widanalage, D. Gaussian Process Regression with Automatic Relevance Determination Kernel for Calendar Aging Prediction of Lithium-ion Batteries. *IEEE Trans. Ind. Inform.* **2019**. [CrossRef]
34. Berckmans, G.; Messagie, M.; Smekens, J.; Omar, N.; Vanhaverbeke, L.; Van Mierlo, J. Cost Projection of State of the Art Lithium-Ion Batteries for Electric Vehicles Up to 2030. *Energies* **2017**, *10*, 1314. [CrossRef]

Article

Machine Learning Techniques for Improving Self-Consumption in Renewable Energy Communities

Zacharie De Grève [1,*], Jérémie Bottieau [1], David Vangulick [2], Aurélien Wautier [1], Pierre-David Dapoz [1], Adriano Arrigo [1], Jean-François Toubeau [1] and François Vallée [1]

1 Power Systems and Markets Research Group, University of Mons, 7000 Mons, Belgium; jeremie.bottieau@umons.ac.be (J.B.); aurlien.wautier@umons.ac.be (A.W.); pierre-david.dapoz@umons.ac.be (P.-D.D.); adriano.arrigo@umons.ac.be (A.A.); jean-francois.toubeau@umons.ac.be (J.-F.T.); francois.vallee@umons.ac.be (F.V.)

2 ORES, Opérateur des Réseaux gaz et électricité, 1348 Louvain-la-Neuve, Belgium; david.vangulick@ores.be

* Correspondence: zacharie.degreve@umons.ac.be

Received: 15 July 2020; Accepted: 14 September 2020; Published: 18 September 2020

Abstract: Renewable Energy Communities consist in an emerging decentralized market mechanism which allows local energy exchanges between end-users, bypassing the traditional wholesale/retail market structure. In that configuration, local consumers and prosumers gather in communities and can either cooperate or compete towards a common objective, such as the minimization of the electricity costs and/or the minimization of greenhouse gas emissions for instance. This paper proposes data analytics modules which aim at helping the community members to schedule the usage of their resources (generation and consumption) in order to minimize their electricity bill. A day-ahead local wind power forecasting algorithm, which relies on state-of-the-art Machine Learning techniques currently used in worldwide forecasting contests, is in that way proposed. We develop furthermore an original method to improve the performance of neural network forecasting models in presence of abnormal wind power data. A technique for computing representative profiles of the community members electricity consumption is also presented. The proposed techniques are tested and deployed operationally on a pilot Renewable Energy Community established on an Medium Voltage network in Belgium, involving 2.25 MW of wind and 18 Small and Medium Enterprises who had the possibility to freely access the results of the developed data modules by connecting to a dedicated web platform. We first show that our method for dealing with abnormal wind power data improves the forecasting accuracy by 10% in terms of Root Mean Square Error. The impact of the developed data modules on the consumption behaviour of the community members is then quantified, by analyzing the evolution of their monthly self-consumption and self-sufficiency during the pilot. No significant changes in the members behaviour, in relation with the information provided by the models, were observed in the recorded data. The pilot was however perturbed by the COVID-19 crisis which had a significant impact on the economic activity of the involved companies. We conclude by providing recommendations for the future set up of similar communities.

Keywords: energy communities; machine learning; forecasting; abnormal data; wind power; outliers; electricity consumption representative profiles; self-consumption

---

1. Introduction

1.1. Context

The operation and planning of modern electric power systems face major transformations nowadays, due to the increasing share of renewable generation (e.g., wind or solar) in the electricity mix, which is uncertain by nature and tends to be deployed in a decentralized way, and to the liberalization and unbundling of the electricity supply chain which occurred in the 1990s in Europe.

The main challenge with electricity systems consists in the fact that generation and consumption must be physically equal at every instant in order to maintain system stability, since electrical energy is as for now difficultly storable at a large scale. Extreme problems of coordination must thereby be solved by modern Transmission System Operators (TSOs), which are furthermore complicated by the fact that they do not own the generation (and consumption) assets, since the liberalization of the electricity sector. The coordination is in that context performed through market platforms on which the market actors can interact. The wholesale market level allows in that way interactions between large producers, large consumers and entities known as Access Responsible Parties (ARPs) (or Balance Repsonible Parties—BRPs—depending on the country), which are responsible for maintaining the balance in their portfolio (containing injection, offtakes and possibly exchanges with other ARPs/BRPs). Bilateral contracts, and power exchange platforms such as EPEX SPOT [1] in Western Europe, provide such opportunities at the wholesale level, with exchange horizons starting from years ahead to close to real time. The retail market enables on the other hand interactions between small end-users (consumers and prosumers) and electricity suppliers (through e.g., fix and varying tariff contracts), which are often themselves ARPs/BRPs.

Currently, new modes of exchange of electricity tend to emerge at the local level, which question the market structure depicted above. This is motivated firstly by the proliferation of decentralized renewable energy resources (owned by small end-users or prosumers), following the ambitious environmental targets promoted at the European and worldwide scale, for which a more efficient coordination could be achieved locally. The increasing willingness of the citizen to play an active role in the electricity supply chain is another important driving factor. The literature speaks of 'consumer-centric electric systems', for which the end-user is placed at the centre of the electrical energy value chain.

Some studies propose in that way to keep a centralized market structure, while adapting the wholesale markets to extend their conditions of access to small end-users [2,3]. On the other hand, fully decentralized structures relying on peer-to-peer exchanges, in which all prosumers and consumers are directly interconnected between each other for buying and selling energy services, are discussed in [4,5]. An intermediate solution promotes the grouping of local consumers/prosumers into organized communities, in which energy resources are pooled and allocated to reach a common objective. The modes of exchanges of energy inside the community may however vary depending on authors: local competitive markets are for instance established in [6,7], whereas collaboration prevails over competition in [8,9]. Peer-to-peer exchanges inside communities are also studied in [10,11].

In its directive 2018/2001 on the promotion of the use of energy from renewable sources [12], the European (EU) Commission has formalized the concept of Renewable Energy Communities (or RECs), in which end-users would be allowed to exchange renewable energy produced locally. The directive has since been transposed into decrees and legal frameworks in many countries of the European Union, e.g., in Wallonia in Belgium [13], in France [14] or in Italy [15]. The science and technology communities have in parallel launched many initiatives to study and implement pilot projects of RECs: the cVPP project [16], lead by the Technische Universiteit of Eindhoven, and the E-Cloud project [17,18], lead by ORES (one of the main Walloon Distribution System Operator or DSO) and which will further be described in Section 3, are two striking examples. The present paper focuses more particularly on the case of such RECs.

1.2. Related Work

Many challenges related to the modeling of RECs are still investigated in the literature, which mainly deal with the optimal operation (e.g., how should we allocate in day-ahead energy resources among members in a community to fulfill a given objective?) and sizing (how should we dimension renewable generation, storage, etc. in a community?) of the communities (see references [6–11] exposed above). More particularly, the multi-agent nature of the underlying optimization problems has driven an increasing attention of the researchers towards game theoretical models for studying the economic equilibria that can appear inside the communities [11,19,20]. Some authors are focusing on the other hand on regulatory aspects related to RECs: these communities consist indeed in a new market design which can play a role at the macroeconomic scale in case of a general adoption. Authors in [21] use cooperative game theory to show for instance that inadequate grid tariffs may lead to an excess adoption of the model, with a potential snowball effect.

Optimally allocating resources in day-ahead in a community requires however to be informed with accurate prospects in terms of local injections (renewable energy production) occurring in the community, with a small time granularity (e.g., quarter hourly). The role of demand (i.e., electricity consumption) response for better matching the generation, in the context of RECs and more generally in electrical power systems, is furthermore well-kown and heavily investigated in the literature, through e.g., the direct control of appliances (see e.g., [22]), or appropriate ex ante recommendations on the consumption behavior of end-users, provided possibly by optimization routines driven by economic signals (see e.g., [23]).

More particularly, data analytics techniques, and especially Machine Learning, can play an important role in better anticipating the generation and demand primitives in communities. The 1 h ahead forecasts of the electricity consumption are for instance performed in [24] using neural networks, in order to support a fuzzy-logic based controller which implements the resource matching in rural communities. Authors in [25] developed a Markov Chain for forecasting a day ahead the aggregated solar generation surplus and residual load in a community comprising storage. A Long Short Term memory network is proposed in [26] to forecast in day-ahead the energy demand in a whole P2P community. Other researchers try on the other hand to avoid the complexity of forecasting models by developing online optimization methods [27,28]. Finally, some studies leverage data analytics for improving the sizing of the communities, such as [29], in which a load profile generator based on Self Organized Maps (SOMs) is proposed.

1.3. Objectives and Contributions

In this paper, we focus on the day-ahead forecasting of time series of local wind power generation in a community, whereas most of the literature studies communities with solar generation only, and on the modeling of the electricity consumption of the individual community members, whereas most of references focus on the consumption quantities aggregated at the community level. We develop data analytics modules, relying on state-of-the-art Machine Learning models, which are expected to help the community members to adapt their consumption profiles to the local renewable energy generation, thereby improving the local coordination. More particularly:

1. we develop a day-ahead local wind power forecasting model, based on the use of state-of-the-art Machine Learning models (tree-based techniques and neural network architectures) trained using a backtesting procedure commonly used in the field of time series forecasting, among the best currently used in worldwide energy forecasting contests [30],
2. we propose an original method for improving the performance of neural network forecasting models in presence of wind power abnormal data, which is quite abundant when performing localized wind power predictions,
3. rather than developing pure load forecasting models for each individual, which is a complex task requiring explanatory variables difficultly obtainable in practice (e.g., for privacy concerns), we propose an algorithm for generating representative profiles of the community members electricity consumption at the considered time of the year, which is solely based on past consumption data,
4. we deploy operationally the developed data analytics modules in a pilot REC established on an industrial area in Tournai (Belgium) on the existing Medium Voltage (MV) distribution grid in the framework of the E-Cloud project [17,18], comprising 2.25 MW of wind generation and

18 Small and Medium Enterprises (SMEs), who had the possibility to freely access the results of the developed data modules by connecting to a dedicated web platform,

5. we quantify the impact of the modules on the operation of the REC (forecasting performance of the developed models, and behaviour of the community members via the evolution of their self-sufficiency and self-consumption during the pilot).

The paper is organized as follows. Section 2 describes the developed data analytics models, with an emphasis on local wind power forecasting in Section 2.1 and on the generation of representative electricity consumption profiles in Section 2.2. Section 3 first describes the pilot REC on which the developed modules are applied (Section 3.1), focuses then on the performance of the wind power forecasting module (Sections 3.2 and 3.3), and finally quantifies the impact of the data analytics modules on the behaviour of the community members and on the operation of the REC, by analyzing the evolution of the community self-consumption and self-sufficiency (Section 3.4).

2. Methodology

The developed data analytics modules, which aim at improving the operation of the RECs through a better coordination between local generation and consumption, are explained in the present section. More particularly, a local day-ahead wind power forecasting model, able to deal with wind power abnormal data, is presented in Section 2.1. Section 2.2 describes the generator of electricity consumption representative profiles.

2.1. *Local Wind Power Forecasting*

This section first describes the different Machine Learning models that are employed for the day-ahead prediction of wind power time series. An original methodology for automatically dealing with abnormal wind power data, which are abundant in the case of localized predictions (as opposed to the case of aggregate predictions made at the regional or national level), during the learning phase of neural network models, thereby improving the forecast performance, is then presented.

2.1.1. Forecasting Model

The forecasting of wind power time series is cast as a Machine Learning regression problem, a particular class of supervized learning problems for which the output is continuous. In this paper, five different models are employed and compared. We first implement two state-of-the-art neural network architectures, namely the traditional feedforward MultiLayer Perceptron or MLP [31], as well as recurrent extensions such as the LongShort Term Memory network or LSTM [32] and its bidirectional variant BLSTM [33], recently applied in the energy sector in [34,35]. We then implement two tree-based techniques, namely Random Forests or RF [36], and Gradient Boosting Decision Tree techniques or GBDT [37]. Finally, the fifth forecasting model (ENSEMBLE) is an ensemble forecast whose output is simply the average of the four previous models. The employed models represent a snapshot of current state-of-the-art Machine Learning techniques, as exemplified by their high performances in contests such as the Global Energy Forecasting Competition [30].

We focus on the day-ahead prediction of wind power, i.e., we aim at forecasting at 12:00 p.m. of day $D-1$ the wind power for the 96 quarters of an hour of day D. Figure 1 depicts the overall procedure. Input features are composed of historical data (i.e., past wind power production and meteorological data such as wind speed, temperature and pression) and of future data (in this case the publicly available day-ahead onshore wind power forecast made by the Transmission System Operator—or TSO—at the national level). In the case of the tree-based models, i.e., RF and GBDT, one model is trained for each of the 96 quarters of an hour of day D, and the quarter hourly forecasts $\mathrm{WP}_q^{\mathrm{RF}}$ (or $\mathrm{WP}_q^{\mathrm{GBDT}}$) , $q = 0, \ldots, 95$ are combined in a forecast vector $\mathbf{WP}_{0-95}^{\mathrm{RF}}$ (or $\mathbf{WP}_{0-95}^{\mathrm{GBDT}}$) for the whole day D. One single MLP model with an output layer of size 96 predicts the wind power for day D $\mathbf{WP}_{0-95}^{\mathrm{MLP}}$, and one single unrolled BLSTM model predicts $\mathbf{WP}_{0-95}^{\text{BLSTM}}$. Finally, the output of the ENSEMBLE model, i.e., $\mathbf{WP}_{0-95}^{\text{Average}}$, is computed by calculating the average of $\mathbf{WP}_{0-95}^{\text{RF}}$, $\mathbf{WP}_{0-95}^{\text{GBDT}}$, $\mathbf{WP}_{0-95}^{\text{MLP}}$ and $\mathbf{WP}_{0-95}^{\text{BLSTM}}$.

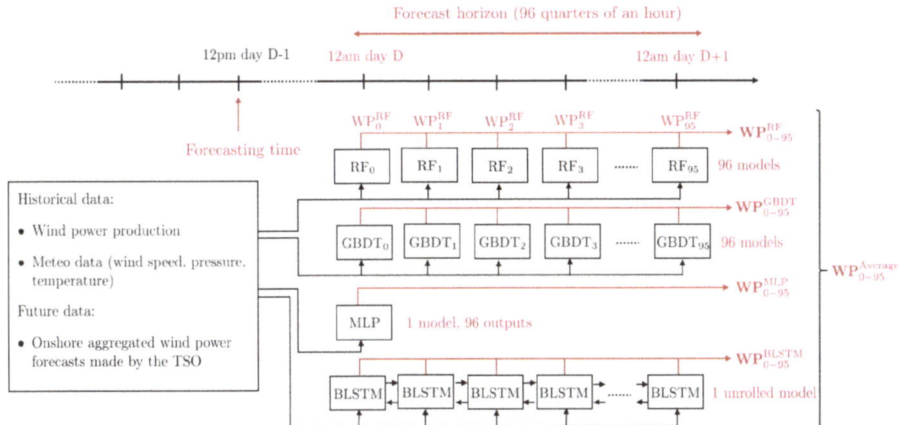

Figure 1. Wind power forecasting methodology.

2.1.2. Strategy for Dealing with Abnormal Wind Power Data

Abnormal data—or outliers—are quite common in localized wind power data, which may have a strong impact on the performance of wind power forecasting models which are built on such data. These can be detected by analyzing the wind turbine power curve (which depicts wind power as a function of wind speed), and can be classified into four categories depending on their position with respect to the normal power curve, according to [38]: bottom curve stacked outliers (due to turbine failure, communication equipment failure, measurement terminal failure, unplanned maintenance—see zone 1 of Figure 2), mid-curve stacked outliers (caused by wind curtailment or communication issues—see zone 2 of Figure 2), top-curve stacked outliers (caused by communication error or wind speed sensor failure—see zone 3 of Figure 2) and around-curve stacked outliers (due to random factors such as signal propagation noise and extreme weather conditions—see zone 4 of Figure 2).

In this paper, we propose an original method for taking into account the presence of abnormal wind power data directly in the learning procedure of the neural network wind power forecast models, in order to improve the forecast performance. In practice (and voluntarily summarizing the process for the sake of clarity), the learning procedure for neural networks, and more generally for supervised learning models, consists in identifying the values of the model parameters $\boldsymbol{\theta}$ (e.g., the weights of a neural network) minimizing a loss function $\mathcal{L}$, which quantifies how well the model fits the training data:

$$\boldsymbol{\theta}^* = \arg\min_{\boldsymbol{\theta}} \mathcal{L}(\hat{\boldsymbol{y}} = f_{\boldsymbol{\theta}}(\boldsymbol{x}), \boldsymbol{y}) \quad (1)$$

with $\hat{\boldsymbol{y}}$ the output of model $f_{\boldsymbol{\theta}}(\boldsymbol{x})$, $\boldsymbol{x}$ the vector of input features, $\boldsymbol{y}$ the target vector (i.e., the true forecast values extracted from the training set $(\boldsymbol{x}_i, \boldsymbol{y}_i), i = 1, \ldots, N$, with N the number of samples in the training set), and $\boldsymbol{\theta}^*$ the optimal parameters values. Problem (1) is solved using variants of the gradient descent algorithm, for most of supervised learning models.

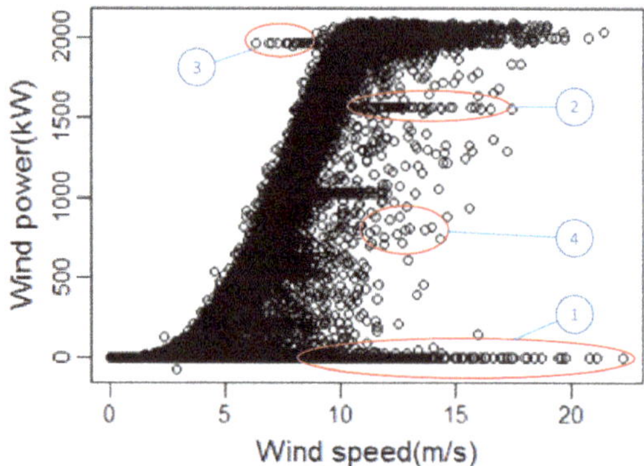

Figure 2. Abnormal data in wind power curves (extracted from [38]).

In that context, the main idea of our adapted learning procedure consists in modifying the loss function $\mathcal{L}$ in order to cancel the contribution of data objects which are tagged as abnormal by an ad hoc abnormal data detection algorithm. The general procedure, depicted in Figure 3, can be described as follows in the case of a neural network model. Each time a training sample $(\boldsymbol{x}_i, \boldsymbol{y}_i)$ is presented at the input of the model, apply the following steps:

1. **Forward pass**. Compute $\hat{\boldsymbol{y}}_i = f_{\boldsymbol{\theta}}(\boldsymbol{x}_i)$, i.e., an estimation of the true forecast $\boldsymbol{y}_i$.
2. **Abnormal data detection**. Detect abnormal wind power data in the target vector $\boldsymbol{y}_i$ using an ad-hoc data detection algorithm. Reference [38] proposes for instance a two-step algorithm, based on a combination of the changing-point grouping method and the quartile method, for automatically detecting and tagging wind power abnormal data, and that we propose to use in the present work. By doing so, a masking vector $\boldsymbol{m}_i$, which contains 1 when the data are normal and 0 if the data are considered as abnormal, is created.
3. **Compute loss function**. Compute the loss function $\mathcal{L}_i$, excluding the contribution of components tagged as abnormal data. The classical L_2-norm is in that way modified as follows:

$$\mathcal{L}_i = \frac{1}{2}\|\boldsymbol{m}_i(\boldsymbol{y}_i - \hat{\boldsymbol{y}}_i)\|_2^2 \tag{2}$$

4. **Backward pass.** Update the parameters (i.e., the weights $\mathbf{W} = \{w_{ij}^l\}$ of the neural network, with $l = 1, \dots, N_L$, $i = 1, \dots, n_{l-1}$, $j = 1, \dots, n_l$, and with N_L the number of layers in the neural network and n_l the number of neurons in layer l) according to standard backpropagation formula.

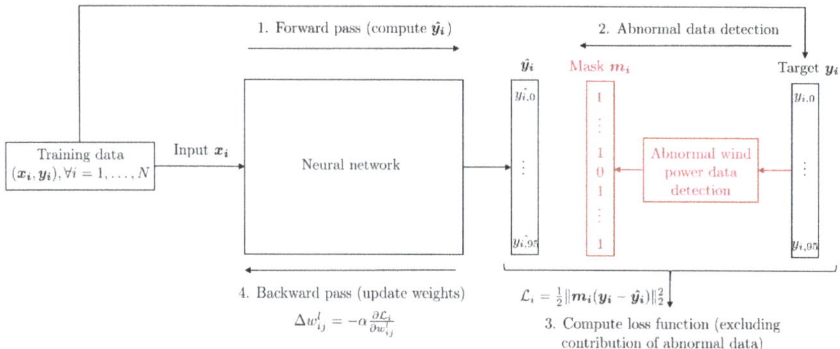

Figure 3. Strategy for training neural networks in presence of wind power abnormal data.

2.2. Electricity Consumption Representative Profiles

Forecasting in day-ahead the electricity consumption of individual companies with the required time granularity (quarter hourly in the present case) is a complex task. The consumption of companies in different branches shows indeed a high variance, as exemplified in [39]. An accurate forecast would require therefore explanatory variables which precisely describe the economic activity of the company, and which are therefore difficultly obtainable in practice, mainly for privacy reasons. In that context, researchers tend to aggregate the electricity consumption at an appropriate level before performing the forecasting task (see e.g., [26,34]), or focus on longer time spans (such as [39] in which the authors predict the annual electricity consumption of enterprises).

In this paper, instead of developing pure forecasting models for each company of the community, we propose a method for generating representative electricity consumption profiles for each member, which is solely based on their past consumption data. The method is inspired by [40] and adapted to the present context.

In the following, we assume that a dataset $\boldsymbol{\mathcal{X}}$ of daily profiles of electricity consumption, sampled at a quarter hourly rate, is available for each member. Each data object x_i is therefore a 96-dimensional vector ($= 4 \times 24$). The procedure is explained below for one community member.

1. Data preprocessing. We firstly remove from the original dataset data objects with missing data.
2. **Segmentation of the full dataset according to calendar information**. For each community member, we segment the available dataset of quarter hourly electricity consumption according to seasons (spring, ..., winter) and days (Monday, ..., Sunday). By doing so, 28 datasets are generated for each member (7 days in one week times 4 seasons). Official off-days are pooled in a separate dataset, so that 29 datasets ($= 28 + 1$) are finally created for each community member.
3. Computing representative profiles for each dataset. Then, a representative profile (or prototype) is calculated for each of the 29 datasets of each client. The medoid μ of each dataset, i.e., the data object for which the sum of distances to all other objects is minimized, is in that way computed:

$$\mu = \underset{y \in \boldsymbol{\mathcal{X}}=\{x_1,x_2,\ldots,x_N\}}{\arg\min} \sum_{i=1}^{N} d(y, x_i) \quad i = 1, \ldots, N \tag{3}$$

   with $\boldsymbol{\mathcal{X}} = \{x_1, x_2, \ldots, x_N\}$ a dataset of N consumption profiles, and $d(.,.)$ a distance function between two data objects. We use a Dynamic Time Warping (DTW) distance in this work, which is a distance originally developed in the field of speech processing [41] but is now generally employed when comparing time series, and more particularly in shape-based time series clustering [42].
4. Generate consumption profiles between two pre-specified dates. Finally, electricity consumption profiles between day d and day d' are created by 1. generating the sequence of dates between d and d' and 2. assigning to each date the corresponding medoid (winter Monday profile, summer Tuesday profile, off-day profile, etc.).

3. Use Case and Results

The data analytics modules described in Section 2 are applied in this section on a pilot REC established in Belgium in the framework of the E-Cloud project [17]. We begin by describing the selected use-case (Section 3.1), focus then on the performance of the wind power forecasting module (Sections 3.2 and 3.3), and finally quantify the impact of the data analytics modules on the behaviour of the community members and on the operation of the REC, by analyzing the evolution of the members self-consumption and self-sufficiency (Section 3.4).

3.1. Use Case Description

The E-Cloud project [17,18], led by ORES (one of the main Walloon DSOs) in collaboration with local private and public entities (Luminus, IDETA, Siemens, DAPESCO, N-Side and the University of Mons- UMONS-), established a pilot REC in Tournai (Belgium), on an industrial area connected to the Medium Voltage electricity distribution network.

The REC involved 18 members (mainly Small or Medium Enterprises or SMEs) and included 18 MW of wind power generation (of which a portion of only 2.25 MW was allocated to the community, the rest was sold through traditional market processes and wasa therefore out of the scope of the present work), as well as 70 kW of peak photovoltaic generation, owned by third-party investors (the community itself could however own the generation assets, which will be studied in future works). A temporary derogation was granted by the regional Walloon regulator ('Commission Wallonne Pour l'Energie', or CWAPE) in order to apply a tailored pricing scheme inside the community; in that way, community members were allowed to purchase, at an advantageous price, (part of) their electricity consumption directly to the local renewable generation when it was available, bypassing the traditional wholesale-retail market structure and favouring local consumption of the local available generation. A distribution key calculated a priori [18] specified the portion of local renewable energy allocated to each community member every quarter of an hour. For the consumption not covered by the local generation, members were free to establish contracts with suppliers in the classical retail market. In the E-Cloud project, thanks to the data analytics modules presented in this paper, community members were furthermore informed in day-ahead of the prospects in terms of renewable energy production, as well as of their own typical electricity consumption profiles at the concerned time of year. They were in that way incentivized to adapt their consumption to local generation via the preferential tariff which was applied in the community.

The project preparatory phase started in 2017, and the pilot was effectively deployed in Tournai, applying the pricing derogation granted by the regulator, from July 2019 to June 2020. During the pilot life, approximately 7500 MW h were produced by local generation, of which 56% have been consumed locally. The total consumption of the 18 involved companies during the full year of the pilot can be observed on Table 1.

Table 1. Total electricity consumption of the 18 member companies during the pilot (July 2019–June 2020).

Member	Total Cons. [MW h]	Member	Total Cons. [MW h]
1	1.07	10	0.2
2	1.01	11	0.21
3	2.29	12	0.22
4	0.76	13	0.21
5	0.073	14	0.15
6	0.44	15	0.2
7	1.89	16	0.14
8	0.49	17	0.41
9	0.27	18	0.02

3.2. Dealing with Wind Power Abnormal Data

We first demonstrated the efficiency of the original procedure proposed in Section 2.1.2 for dealing with abnormal wind power data in the training of neural network based forecast models. To that end, we leveraged a dataset made available in the framework of the E-Cloud project, consisting of approximately 1.4 years (January 2018–May 2019) of:

- historical wind power data for the wind farm installed in Tournai, sampled at a quarter hourly scale,
- historical meteorological data (wind speed, atmospheric pressure, temperature), sampled at a quarter hourly scale,

- onshore wind power forecasts at the national level, made available publicly by the Belgian Transmission System Operator (TSO) Elia [43] with the objective to benefit market participants and improve the electricity market outcomes [44], sampled at a quarter hourly scale.

The abnormal data detection algorithm [38] presented in Section 2.1.2 was first applied on the farm level wind power curve created using wind power and wind speed data. The outcome of the procedure is depicted in Figure 4, where green circles refer to normal data, and red crosses (blue stars) correspond to abnormal data detected by the quartile method (change point method respectively).

The abnormal data points were employed to create masking vectors $\boldsymbol{m}_i$, which were involved in the modified learning procedure of the wind power neural network forecasting models. A 96-output MultiLayer Perceptron (MLP) which aimed at performing a day-ahead wind power forecast was in that way trained on the dataset depicted above, according to the procedure of Figure 3. More precisely, in accordance with standard text books in Machine Learning [31], a backtesting procedure based on cross-validation and which respected the temporal order of observations, which is common in the field of time series prediction, was performed by decomposing the 1.4 years dataset into three sets: a 13-month training set (January 2018–February 2019) which was used for estimating the model parameters (e.g., the weights of the neural networks, etc.), a 1-month validation set (March 2019) which was employed for tuning model hyperparameters (such as e.g., the number of neurons per layer and the number of layers in neural networks, etc.) and prevent overfitting, as well as a 2-month test set (April–May 2019), for evaluating the model performances on new data that had not been seen previously by the model.

The Python libraries Keras [45] and TensorFlow [46] were employed for implementing and training the neural networks. The Adam optimization algorithm [47], a state-of-the-art variant of stochastic gradient descent, was selected as the training algorithm for estimating the neural network weights. The Tree-structured Parzen Estimator (TPE) approach [48] was employed for optimizing the hyperparameters of the neural network (i.e., the number of hidden layers, the number of neurons in each layer, the size of the input feature vector, etc.), with the help of the *Hyperopt* Python library [49], which led to an MLP architecture with one hidden layer, 32 neurons, and an input layer including 12 past time steps for the wind power, seven past time steps for the wind speed and the atmospheric pressure, and 31 past time steps for the temperature.

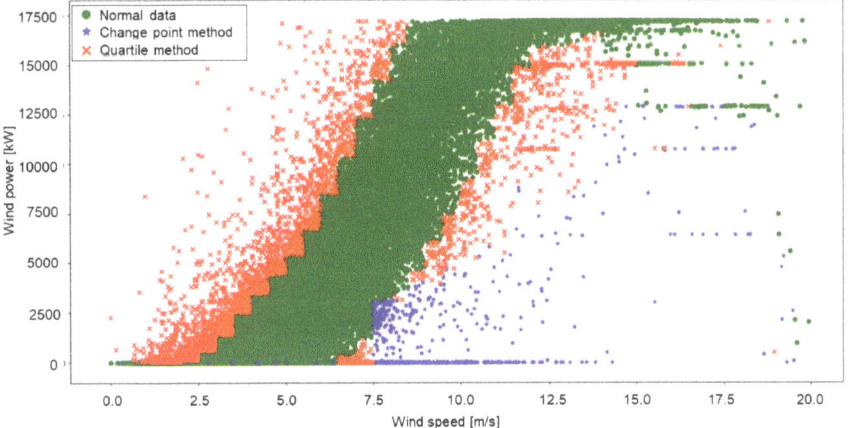

Figure 4. Abnormal wind power data filtering on the E-Cloud data, according to the procedure exposed in [38]. Normal points are tagged with green circles, and abnormal points are tagged with blue stars (red crosses) if they have been identified using the change point method (quartile method).

We then compared the performance of the trained neural network in two different configurations, i.e., when masking the contribution of wind power abnormal data during training according to the procedure exposed in Section 2.1.2 ('MLP with mask'), and without applying any masking effect ('MLP without mask'). To that end, we trained 100 neural networks with the best architecture found above, and computed the Root Mean Square Error (RMSE) obtained on the test set by comparing the forecast and the true value of wind power generation. The Adam training algorithm was indeed a stochastic algorithm, which aimed at minimizing a highly non convex cost function, and which thereby ended up in local minima which varied according to different initial conditions, training parameter values, etc. [47]. The average of the RMSEs, as well as the standard deviation, the min and max values of the RMSE, are depicted in Table 2, for the two approaches (with and without mask). It is shown that the masking of wind power abnormal data was able to decerase the average RMSE by approximately 10%, which confirms the interest and efficiency of the proposed methodology.

Table 2. Strategy for dealing with wind power abnormal data: forecasting performance of the trained MultiLayer Perceptrons (MLPs) with and without the proposed mask.

	MLP without Mask	MLP with Mask
μ_{RMSE} [kW]	2634.8	2436.2
σ_{RMSE} [kW]	98.7	58.8
max RMSE [kW]	2892.9	2621.5
min RMSE [kW]	2414.9	2337.7

3.3. Forecasting Performance

The performance of the five day-ahead wind power forecasting models described in Section 2.1.1 is studied in this section. The neural networks models were implemented in Keras [45], using the original training procedure presented in the previous section, and the tree-based models (RF and GBDT) were implemented in Python using the Scikitlearn library [50]. The output of the ENSEMBLE model was simply coded in Python by computing the average of the outputs of the four other models. The same dataset and input features than the previous section were employed, and cross-validation was also performed for the training-evaluation procedure. Figure 5 depicts the wind power forecast obtained with the ENSEMBLE model (in red) as a function of time, as well as the actual wind power generation (in black), for a random day of the test set. One can observe that, even if the forecast error remained clearly visible, the model was most of the time able to correctly capture the time of day when the peak of wind power generation occurred, which is fundamental information for the community members for scheduling their consumption for the upcoming day.

Table 3 shows the RMSE of the five developed forecast models. We observed that the RF model was the individual model which provided the smallest forecast error in this particular application, and that the ENSEMBLE model (which was simply built by taking the average of the four other models), was still able to slightly improve the accuracy. The ENSEMBLE model was therefore selected for the operational deployment described in Section 3.4.

Table 3. Forecasting performance in terms of Root Mean Square Error (RMSE), for the five local day-ahead wind power forecast models (Random Forest—RF, Gradient Boosting Decision Tree—GBDT, MultiLayer Percpetron—MLP, Bi-LSTM—BLSTM, and ENSEMBLE.

	RF	GBDT	MLP	Bi-LSTM	Ensemble (Average)
RMSE [kW]	2347	2387	2338	2389	2327

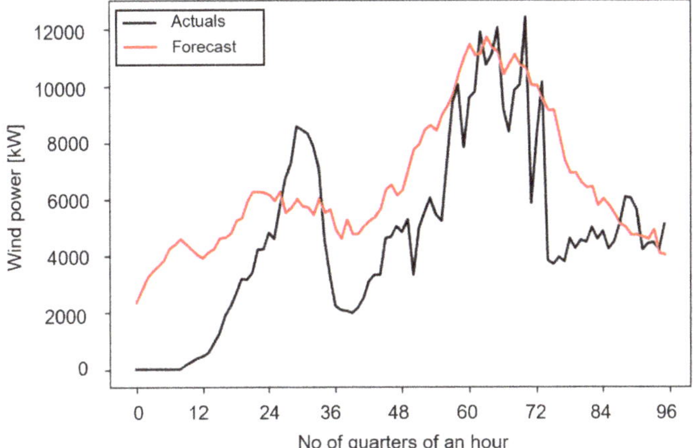

Figure 5. Time series of wind power forecast and actuals for a day of the test set.

3.4. Impact on the Consumption Behaviour of the Renewable Energy Community Members

The two data analytics modules developed in this work were deployed operationally in the E-Cloud pilot using Mindsphere [51], a cloud-based open IoT operating system developed by Siemens. Each member of the pilot community received a personal access that he used to freely connect to a dedicated web platform using his personal computer, on his own initiative. Each member was in that way able to consult general data such as his own monthly self-consumption or self-sufficiency (see definitions below), as well as the same quantities for the whole community. It is important to mention that personal data from the other community members was hidden, for the sake of privacy. As explained in the previous sections, a common (i.e., the same for each member) day-ahead renewable generation forecast under the form of a quarter hourly time series, which was refreshed every day at 12 pm, was also made available to each member, using the methodology of Section 2.1. Each member was also able to consult a typical consumption profile for the upcoming day, representative of his past consumption at the considered time of year, according to the procedure exposed in Section 2.2. Given the preferential tariff that was in application in the community for the purchase of energy which was locally produced, we expected that the members would take advantage of the information provided by the data modules, on their own initiative, in order to adapt their consumption profiles to local generation, thereby decreasing their energy bill.

The modules became effectively operational from April 2020 to June 2020. A downscaled version of the regional solar forecast made available publicly by the TSO Elia [43] was employed for the 70 kW of solar generation installed in the community, since the absence of metered solar data in the pilot prevented the training of a local solar forecast model. The amount of installed wind generation—1.8MW—was however 20 times higher than installed PV power, which mitigated the necessity to have a very accurate solar forecasting module in this particular case.

We show the impact of the data analytics modules on the behaviour of the community members by computing the monthly self-consumption of members, i.e., the ratio between the member self-consumed energy (i.e., the member electrical energy consumption covered by the local energy which was put at his disposal) and the local energy which was put at his disposal (i.e., the portion of local generation that was allocated to him, according to predefined distribution keys mentioned in Section 3.1), during one month. This first index quantified to what extent the community generation tended to be consumed locally, where it has been produced: a self-consumption of 100% for a member meant for instance that he had consumed all the local renewable generation that was allocated to him for the considered month. The monthly self-consumption $SC_{i,m}$ of member i during month m was in that way expressed as follows:

$$SC_{i,m} = \frac{E_{\text{self},i,m}}{\eta_i E_{\text{gen},m}}, \forall i \in \mathcal{I}, m \in \mathcal{M}, \tag{4}$$

with $E_{\text{gen},m}$ the total energy generated locally in the community during month m, η_i the fraction of that energy that is allocated to member i (constant during the whole pilot), $E_{\text{self},i,m}$ the energy consumed by member i during month m that was covered by $\eta_i E_{\text{gen},m}$, and $\mathcal{I}$ ($\mathcal{M}$) the set of community members i (respectively considered months m). It should be noted that if renewable energy was allocated to a member who did not consume it entirely, the corresponding excess of energy was not counted in the numerator of (4).

Similarly, we computed the monthly self-sufficiency of members, i.e., the ratio between the member electrical energy consumption covered by the local generation that was allocated to him and the total energy consumed by the member, again during 1 month. This index shows what part of his electricity consumption the member consumed from local resources, and by extension what part he had to purchase on the traditional retail market: a self-sufficiency of 100% means that the member was able to cover all its consumption with the local generation that was allocated to him during the considered month. Self-sufficiency of member i during month m is in that way computed as follows:

$$SS_{i,m} = \frac{E_{\text{self},i,m}}{E_{\text{cons},i,m}}, \forall i \in \mathcal{I}, m \in \mathcal{M}, \tag{5}$$

with $E_{\text{cons},i,m}$ the total energy consumed by member i during month m.

The left part of Figure 6 shows the monthly self-consumption of the 18 community members during the pilot duration, i.e., from July 2019 to June 2020. It is first very important to note that the data analytics module were effectively deployed on-site in April 2020, one month after the generalized lockdown that occurred in Belgium as a consequence of the COVID-19 crisis. The economic activity of the 18 companies involved in the community suffered in that context from a drastic reduction, which has been materialized by a significant drop of their electricity consumption, while the generation remained unchanged compared to the pre-COVID situation. This explains in the authors opinion why the self-consumption of almost all members significantly decreased starting from March 2020 to May 2020, with a progressive increase in May and June 2020, in line with the progressive removal of lockdown measures that occured in mid-May 2020 in Belgium. This effect masked unfortunately the possible positive impact of the data analytics modules on the behaviour of the community members.

The right part of Figure 6 depicts the monthly self-sufficiency of each member during the pilot duration, which should be a priori less impacted by the COVID-19 crisis since it is a ratio between two consumptions, namely the member consumption covered by local resources and the member total consumption, which are both expected to decrease in the COVID-19 situation. We observed a global increasing trend in the self-sufficiency of the community members from March to June 2020. It was however not possible to entirely attribute this positive effect to the operational availability of the developed data modules: the effect of the COVID-19 crisis on the trends in self-sufficiency could not be completely discarded, since changes in economic activity may have modified the shape of the daily consumption patterns (due to the temporary suspension of some industrial processes, etc.), which can impact the self-sufficiency as likely. Furthermore, the global increase in terms of self-sufficiency may also be attributed to a yearly seasonal effect, which is possible considering the values in July 2019, at the beginning of the pilot. The time span covered by the pilot, i.e., one year according to the special derogation granted by the Walloon regulator (CWAPE), is however not sufficient to discard or confirm that hypothesis.

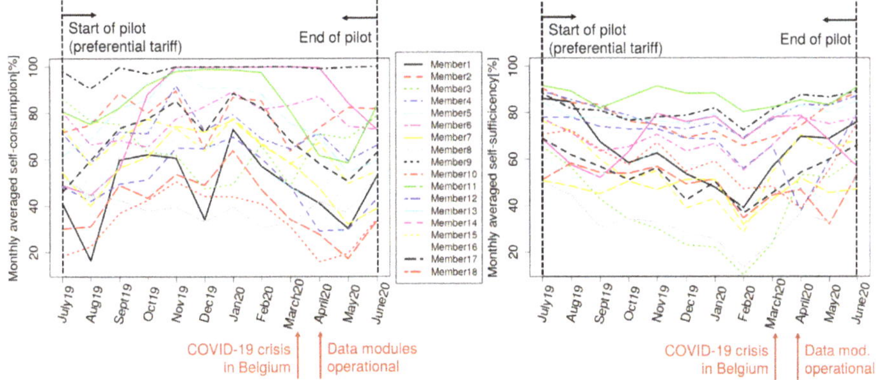

Figure 6. Monthly self consumption (**left**) and self-sufficiency (**right**) of the 18 members of the E-Cloud pilot. The preferential tariff became effective in July 2019, which started the pilot officially, and ended late June 2020. The data analytics modules developed in this work were effectively deployed on site in April 2020, during approximately 3 months.

Finally, we show in Table 4 the relative change in terms of self-sufficiency for each member between July 2019 and June 2020, in percent. We expect in that way to compare consumer habits at almost one year interval, which can be a better indicator of possible changes in consumption patterns. For 11 out of the 18 members, it appeared that the self-sufficiency decreased, whereas it increased for six members. Again, no significant impact of the data modules on the consumption behaviour was observed (to be more conclusive, July 2020 should have been compared with July 2019, but the pilot was scheduled to end in June 2020 as explained above). The effect of the COVID-19 crisis in June 2020 could not be completely discarded as well, since the economic activity in Belgium had not recovered its pre-COVID intensity in July 2020 yet, at the time of writing.

Table 4. Change ϵ_{SS} in self-sufficiency between July 2019 and June 2020 for each community member, in percent.

Member	1	2	3	4	5	6	7	8	9	10	11	12	13	14	15	16	17	18
ϵ_{SS} [%]	−10	−2	−5	−2	0.5	−12	−8	−5	1	2	−0.8	1.5	−9.8	0.4	−3.7	NA	−3	4.5

4. Conclusions and Perspectives

This work proposed energy analytics tools to inform the members of Renewable Energy Communities (RECs) of the day-ahead prospects in terms of local renewable energy generation, as well as in terms of electricity consumption profiles which are representative of the members behaviour at the considered time of the year. By doing so, the members were expected to adapt their own consumption patterns to local generation, in order to benefit of advantageous energy pricing mechanisms which prevail in a community.

A localized day-ahead wind power forecasting tool, based on state-of-the-art Machine Learning algorithms, has been developed in that way. The ENSEMBLE model, whose output is computed as the average of the outputs of four other Machine Learning models (Random Forests, Gradient Boosting Decision Trees, a MultiLayer Perceptron and a Bi-directional LSTM) has shown the best forecasting performance on the E-Cloud pilot project data. Forecasting accuracy has been further improved by automatically detecting wind power abnormal data samples and by adapting the training procedure accordingly. A procedure for generating representative electricity consumption profiles of the community members, relying on Dynamic Time Warping (a state-of-the-art Machine Learning distance employed when comparing time series), has further been implemented.

The data modules have been deployed on-site in the framework of the E-Cloud pilot project, on a REC connected to the existing Medium Voltage distribution grid in an industrial area in Belgium, and composed by 18 members (mainly Small and Medium Enterprises or SMEs) and by local generation (mainly wind power). The information provided by the data modules was freely available to the community members by connecting to a dedicated web platform on their own initiative. Global quantities, such as the monthly self-sufficiency and self-consumption of the community members, have been computed to quantify the impact of the data modules on the consumption behaviour of the community members. We were not able however to highlight significant changes in the consumer habits. It is worth mentioning though that the general lockdown that occurred in Belgium in March 2020 due to the COVID-19 crisis significantly affected the results, especially knowing that the data modules became operational for the first time in April 2020, during lockdown. Yearly seasonal effects were furthermore observed in the self-sufficiency patterns, which further masked the potential benefits of the deployed data modules. We recommend therefore to extend in accordance with the local regulator the duration of similar REC pilots to more than one year, in order to better understand these yearly seasonal effects, and to better quantify the impact on the members consumption behaviour. Furthermore, we strongly encourage researchers and industrials that will implement similar pilot RECs in the future to establish a system for monitoring the usage of the displayed information by the community members (for instance by recording the number of connections to the dedicated web platform), in order to quantify and possibly stimulate their interest in the provided tools.

As a first perspective, we intend to deploy a pilot REC for a longer time span, with an active monitoring of the members interest in the available tools, in order to confirm/infirm the hypotheses raised in this work. This is however a slow process, since temporary derogations by the local regulator are mandatory currently in Belgium for applying a community-based pricing scheme. We further aim at building another pilot REC in a residential area, in order to analyze the behaviour of domestic consumers. We finally intend to improve the accuracy of the wind power forecasting module by using turbine-level data in the model definition, and by adapting the learning procedure of tree-based algorithms to the presence of wind power abnormal data. We also intend to focus our research effort on the correct prediction of peaks of generation, since the community benefits are optimized when members shift their consumption to generation peak times, and on the recalibration of the wind power forecasting models in the flavour of [52].

Author Contributions: Conceptualization, Z.D.G. and D.V.; Data curation, J.B. and A.A.; Methodology, Z.D.G., J.B., A.W., P.-D.D. and J.-F.T.; Project administration, Z.D.G. and D.V.; Resources, D.V. and F.V.; Supervision, F.V.; Validation, D.V. and A.A.; Writing—original draft, Z.D.G.; Writing—review & editing, J.B., D.V., A.A., J.-F.T. and F.V.. All authors have read and agreed to the published version of the manuscript.

Funding: This research has been funded by ORES. J.-F. Toubeau is supported by F.R.S./FNRS (Belgian National Fund of Scientific Research).

Acknowledgments: The authors would like to thank the participants of the E-Cloud project for their valuable inputs to the present study: ORES, Luminus, IDETA, Siemens, N-Side, DAPESCO.

Conflicts of Interest: The authors declare no conflict of interest.

References

1. EPEX Spot. Available online: https://www.epexspot.com/en (accessed on 10 July 2020).
2. Elia Group. Towards a Consumer-Centric Electric System. 2018. Available online: https://www.elia.be/-/media/project/elia/elia-site/company/publication/studies-and-reports/studies/elia-vision-paper-2018_front-spreads-back.pdf (accessed on 15 Feburary 2020).
3. Pinson, P.; Baroche, T.; Moret, F.; Sousa, T.; Sorin, E.; You, Z. The Emergence of Consumer-Centric Electric Markets. 2018. Available online: http://pierrepinson.com/docs/pinsonetal17consumercentric.pdf (accessed on 10 January 2020).
4. Parag, Y.; Sovacool, B.K. Electrcity market design for the prosumer era. *Nat. Energy* **2016**, *1*, 16032. [CrossRef]

5. Crosby, M. An Airbnb or Uber for the Electricity Grid? How DERs Prepare the Power Sector to Evolve into a Sharing Economy Platform. RMI Outlet Blog: Rocky Mountain Institute. Available online: https://rmi.org/airbnb-uber-electricity-grid/ (accessed on 10 July 2020).
6. Moret, F.; Pinson, P. Energy Collectives: A Community and Fairness Approach to Future Electricity markets. *IEEE Trans. Power Syst.* **2019**, *34*, 3994–4004. [CrossRef]
7. Duschene, L.; Cornélusse, B.; Savelli, I. Sensitivity analysis of a local market model for community microgrids. In Proceedings of the PowerTech Conference (PowerTech2019), Milan, Italy, 23–27 June 2019.
8. Hupez, M.; De Grève, Z.; Vallée, F. Cooperative demand-side management scenario for the low-voltage network in liberalised electricity markets. *IET Gener. Transm. Distrib.* **2018**, *12*, 5990–5999. [CrossRef]
9. Hupez, M.; Toubeau, J.-F.; De Grève, Z.; Vallée, F. A New Cooperative Framework for a Fair and Cost-Optimal Allocation of Resources within a Low Voltage Electricity Community. *IEEE Trans. Smart Grids*, Under Review.
10. Morstyn, T.; McCulloch, M. Multiclass energy management for peer-to-peer energy trading driven by prosumer preferences. *IEEE Trans. Power Syst.* **2019**, *34*, 2005–2014. [CrossRef]
11. Cadre, H.L.; Jacquot, P.; Wan, C.; Alasseur, C. Peer-to-peer electricity market analysis: from variational to Generalized Nash equilibrium. *Eur. J. Oper. Res. (EJOR)* **2020**, *282*, 753–771. [CrossRef]
12. Directive (EU) 2018/2001 of the European Parliament and of the Council of 11 December 2018 on the Promotion of the Use of Energy from Renewable Sources. Available online: https://eur-lex.europa.eu/legal-content/EN/TXT/?uri=uriserv:OJ.L_.2018.328.01.0082.01.ENG&toc=OJ:L:2018:328:TOC (accessed on 5 April 2020).
13. Walloon Parliament-Belgium, Décret modifiant les décrets des 12 avril 2001 relatif à l'organisation du marché régional de l'électricité, du 19 décembre 2002 relatif à l'organisation du marché régional du gaz et du 19 janvier 2017 relatif à la méthodologie Tarifaire Applicable aux Gestionnaires de réseau de Distribution de gaz et d'électricité en vue de Favoriser le déVeloppement des Communautés d'énergie Renouvelable. April 2019. Available online: https://wallex.wallonie.be/contents/acts/19/19007/1.html (accessed on 10 Feburary 2020).
14. French Parliament. Loi du 8 Novembre 2019 Relative à l'énergie et au climat. 8 November 2019. Available online: https://www.vie-publique.fr/loi/23814-loi-energie-et-climat-du-8-novembre-2019 (accessed on 10 February 2020).
15. Gazzetta Ufficiale Della Republica Italiana. Conversione in legge, con modificazioni, del decreto-legge 30 dicembre 2019, n. 162, recante disposizioni urgenti in materia di proroga di termini legislativi, di organizzazione delle pubbliche amministrazioni, nonche di innovazione tecnologica (20G00021). 28 February 2020. Available online: https://www.gazzettaufficiale.it/eli/id/2020/02/29/20G00021/sg (accessed on 10 July 2020).
16. cVPP—Community-Based Virtual Power Plant: A Novel Model of Radical Decarbonisation Based on Empowerment of Low-Carbon Community Driven Energy Initiatives. Interreg NW Europe, 2017–2020. Avilable online: https://www.nweurope.eu/projects/project-search/cvpp-community-based-virtual-power-plant/ (accessed on 15 May 2020).
17. "L'« E-Cloud »: l'autoconsommation collective au service des entreprises, avec des perspectives d'économie sur la facture d'électricité de 8 à 14%", Cluster TWEED, August 2019. Available online: https://clusters.wallonie.be/tweed-fr/08-08-2019-l-e-cloud-l-autoconsommation-collective-au-service-des-entreprises-avec-des-perspectives-d-economie-sur-la-fa.html?IDC=6928&IDD=121138 (accessed on 25 March 2020).
18. Vangulick, D.; Cornélusse, B.; Vanherck, T.; Devolder, O.; Lachi, S. E-CLOUD, the open microgrid in existing network infrastructure. In Proceedings of the 24th International Conference & Exhibition on Electricity Distribution (CIRED2017), Glasgow, UK, 12–15 June 2017; pp. 2703–2716.
19. Jacquot, P.; Beaude, O.; Gaubert, S.; Oujdane, N. Analysis and implementation of an hourly billing mechanism for demand response management. *IEEE Trans. Smart Grids* **2019**, *10*, 4265–4278. [CrossRef]
20. Rahi, G.E.; Etesami, S.R.; Saad, W.; Mandayam, N.; Poor, H.V. Managing price uncertainty in prosumer-centric energy trading: A prospect-theoretic Stackelberg game approach. *IEEE Trans. Smart Grids* **2019**, *10*, 702–713. [CrossRef]
21. Abada, I., Ehrenmann, A.; Lambin, X. Unintended consequences, the snowball effect of energy communities. *Energy Policy* **2020**, *143*, 14. [CrossRef]

22. Conejo, A.; Morales, J.M.; Baringo, L. Real time demand response model. *IEEE Trans. Smart Grids* **2010**, *1*, 236–242. [CrossRef]
23. Atzeni, I.; Ordonez, L.G.; Scutari, G.; Palomar, D.P.; Fonollosa, J.R. Demand-side management via distributed energy generation and storage optimization. *IEEE Trans. Smart Grids* **2013**, *4*, 866–876. [CrossRef]
24. Maheshwari, Z.; Ramakumar, R. A framework for intelligent control of SIRES in rural commuities. In Proceedings of the 2017 IEEE Power & Energy Society General Meeting (PESGM), Chicago, IL, USA, 16–20 July 2017. [CrossRef]
25. AlSkaif, T.; Schram, W.; Litjens, G.; van Sark, W. Smart charging of community storage units using Markov chains. In Proceedings of the 2017 IEEE PES Innovative Smart Grid Technologies Conference Europe (ISGT-Europe), Torino, Italy, 26–29 September 2017. [CrossRef]
26. Zou, L.; Munir, M.S.; Kim, K.; Hong, C.S. Day-ahead Energy Sharing Schedule for the P2P Prosumer Community Using LSTM and Swarm Intelligence. In Proceedings of the 2020 International Conference on Information Networking (ICOIN), Barcelona, Spain, 7–10 January 2020. [CrossRef]
27. Jia, K.; Liu, B.; Iyogun, M.; Bi, T. Smart control for battery energy storage system in a community grid. In Proceedings of the 2014 International Conference on Power System Technology, Chengdu, China, 20–22 October 2014. [CrossRef]
28. Wang, G. Lyapunov Optimization Based Online Energy Flow Control for Multi-energy Community Microgrids. In Proceedings of the 2019 IEEE PES GTD Grand International Conference and Exposition Asia (GTD Asia), Bangkok, Thailand, 19–23 March 2019. [CrossRef]
29. Llanos, J.; Saez, D.; Palma-Behnke, R.; Nunez, A.; Jimenez-Estéves, G. Load profile generator and load forecasting for a renewable based microgrid using Self Organizing Maps and neural networks. In Proceedings of the 2012 International Joint Conference on Neural Networks (IJCNN), Brisbane, Australia, 10–15 June 2015. [CrossRef]
30. Hong, T.; Xie, J.; Black, J. Global energy forecasting competition 2017: Hierarchical probabilistic load forecasting. *Int. J. Forecast.* **2019**, *35*, 1389–1399. [CrossRef]
31. Hastie, T.; Tibshirani, R.; Friedman, J. *The Elements of Statistical Learning: Data Mining, Inference, and Prediction*; Springer: New York, NY, USA, 2009. [CrossRef]
32. Hochreiter, S.; Schmidhuber, J. Long Short-Term Memory. *Neural Comput.* **1997**, *9*, 1735–1780.
33. Schuster, M.; Paliwal, K.K. Bidirectional Recurrent Neural Networks. *IEEE Trans. Signal Process.* **1997**, *45*, 2673–2681. [CrossRef] [PubMed]
34. Toubeau, J.-F.; Bottieau, J.; Vallée, F.; De Grève, Z. Deep Learning-based Multivariate Probabilistic Forecasting for Short-Term Scheduling in Power Markets. *IEEE Trans. Power Syst.* **2019**, *34*, 1203–1215. [CrossRef]
35. Toubeau, J.-F.; Bottieau, J.; Vallée, F.; De Grève, Z. Improved day-ahead predictions of load and renewable generation by optimally exploiting multi-scale dependencies. In Proceedings of the 7th IEEE Conf. Innovative Smart Grid Technology, Torino, Italy, 26–29 September 2017. [CrossRef]
36. Breiman, L. Random Forests. *Mach. Learn.* **2001**, *45*, 5–32.
37. Chen, T.; Guestrin, C. XGboost: A scalable tree boosting system. In Proceedings of the 22nd ACM SIGKDD International Conference on Knowledge Discovery and Data Mining, San Francisco, CA, USA, 13–17 August 2016; pp. 785–794. [CrossRef]
38. Shen, X.; Fu, X.; Zhou, C. A combined algorithm for cleaning abnormal data of wind turbine power curve based on change point grouping algorithm and quartile algorithm. *IEEE Trans. Sustain. Energy* **2019**, *10*, 46–54.
39. Stingl, C.; Hopf, K.; Staake, T. Explaining and predicting annual electricity demand of enterprises—A case study from Switzerland. *Energy Inform.* **2018**, *1*. [CrossRef]
40. Aguas, A. EDPD'S experience with data analytics and stochastic simulation methods for risk-controlled network planning. In Proceedings of the 25th International Conference on & Exhibition on Electricity Distribution (CIRED2018), Ljublana, Slovenia, 7–8 June 2018. [CrossRef]
41. Sakoe, H.; Chiba, S. Dynamic programming algorithm optimization for spoken word recognition. *IEEE Trans. Acoust. Speech Signal Process.* **1978**, *26*, 43–49.
42. Aghabozorgi, S.; Shirkhorshidi, A.S.; Wah, T.Y. Time-Series Clustering—A Decade Review. *Inf. Syst.* **2015**, *53*, 16–38. [CrossRef]
43. Available online: https://www.elia.be/fr/donnees-de-reseau/production/donnees-de-production-eolienne (accessed on 10 July 2019). [CrossRef]

44. Exizidis, L.; Kazempour, J.; Pinson, P.; De Grève, Z; Vallée, F. Impact of public aggregate wind forecasts on electricity market outcomes. *IEEE Trans. Sustain. Energy* **2017**, *8*, 1394–1405.
45. Chollet, F. Keras, Github. 2015. Available online: https://github.com/fchollet/keras (accessed on 20 July 2019). [CrossRef]
46. Abadi, M. Tensorflow: A system for large-scale machine learning. In Proceedings of the 12th USENIX Symposium on Operating Systems Design and Implementation (OSDI'16), Savannah, GA, USA, 2–4 November 2016; pp. 265–283.
47. Kingma, D.P.; Ba, J. Adam: A Method for Stochastic Optimization. In Proceedings of the 3rd International Conference for Learning Representations (ICLR2015), San Diego, CA, USA, 7–9 May 2015.
48. Bergstra, J.; Bardenet, R.; Bengio, Y. Algorithms for hyperparameter optimization. In *Advances in Neural Information Processing Systems 24*; Curran Associates Inc.: Dutchess County, NY, USA, 2011; pp. 2546–2554.
49. Bergstra, J.; Komer, B.; Eliasmith, C.; Yamins, D.; Cox, D.D. Hyperopt: A Python library for model selection and hyperparameter optimization. *Comput. Sci. Discov.* **2015**, *8*, 014008.
50. Pedregosa, F. Scikit-learn: Machine Learning in Python. *J. Mach. Learn.* **2011**, *12*, 2825–2830.
51. Siemens. MindSphere, the Cloud-Based, Open IoT Operating System. Available online: https://siemens.mindsphere.io/en (accessed on 10 September 2019). [CrossRef]
52. Toubeau, J.-F.; Dapoz, P.-D.; Bottieau, J.; Wautier, A.; De Grève, Z.; Vallée, F. Recalibration of Recurrent Neural Networks for Short-Term Wind Power Forecasting. *Electr. Power Syst. Res.* **2021**, *190*, 1–7.

Article

Deep Reinforcement Learning-Based Voltage Control to Deal with Model Uncertainties in Distribution Networks

Jean-François Toubeau, Bashir Bakhshideh Zad, Martin Hupez, Zacharie De Grève and François Vallée *

Power Systems and Markets Research Group, University of Mons, 7000 Mons, Belgium; Jean-Francois.TOUBEAU@umons.ac.be (J.-F.T.); Bashir.BAKHSHIDEHZAD@umons.ac.be (B.B.Z.); Martin.HUPEZ@umons.ac.be (M.H.); Zacharie.DEGREVE@umons.ac.be (Z.D.G.)

* Correspondence: Francois.VALLEE@umons.ac.be

Received: 28 June 2020; Accepted: 28 July 2020 ; Published: 1 August 2020

Abstract: This paper addresses the voltage control problem in medium-voltage distribution networks. The objective is to cost-efficiently maintain the voltage profile within a safe range, in presence of uncertainties in both the future working conditions, as well as the physical parameters of the system. Indeed, the voltage profile depends not only on the fluctuating renewable-based power generation and load demand, but also on the physical parameters of the system components. In reality, the characteristics of loads, lines and transformers are subject to complex and dynamic dependencies, which are difficult to model. In such a context, the quality of the control strategy depends on the accuracy of the power flow representation, which requires to capture the non-linear behavior of the power network. Relying on the detailed analytical models (which are still subject to uncertainties) introduces a high computational power that does not comply with the real-time constraint of the voltage control task. To address this issue, while avoiding arbitrary modeling approximations, we leverage a deep reinforcement learning model to ensure an autonomous grid operational control. Outcomes show that the proposed model-free approach offers a promising alternative to find a compromise between calculation time, conservativeness and economic performance.

Keywords: voltage control; deep deterministic policy gradient; deep reinforcement learning; model uncertainties

1. Introduction

The massive integration of Distributed Generation (DG) units in electric distribution networks poses significant challenges for system operators [1–5]. Indeed, distribution networks were historically sized (with a radial structure) to meet maximum load demands while avoiding under-voltages at the end of the lines. However, in presence of local generation, the opposite over-voltage problem may appear. In case of severe voltage violation, inverters of DG units are temporarily cut off. This induces not only a loss of renewable-based energy, but also a deterioration of the delivered power quality (due to resulting voltage and current transients) that accelerates the equipment degradation [6]. In this context, the objective of modern Distribution System Operators (DSOs) is to adopt a reliable and cost-efficient strategy that is able to maintain a safe voltage profile in both normal and contingency conditions, with the goal of enhancing the ability of the system to accommodate new renewable-based resources. To that end, researchers have developed a wide range of techniques, with the aim of avoiding costly investment plans that simply upgrade/reinforce the network. Also, static (experience-based) strategies based on past observations have shown limitations, as they are often sub-optimal and unable to react in a very short time frame (to prevent cascading faults just after a disturbance) [7].

Theoretically, different methods can be applied for voltage management of Medium-Voltage (MV) distribution systems, but the most common methods are based on using on-load tap changer mechanism of the transformer, reactive power compensation and curtailment of DG active powers [8,9]. It is generally known that each of the above voltage control methods has its own advantages and drawbacks, and there is no single perfect voltage regulation method [10]. Recently, there has been a growing literature focusing purely on local strategies in which resources rely only on localized measurements of the voltages' magnitude, and do not exchange information with other agents [11,12]. Such local algorithms are easy to implement in practice, but the lack of a global vision may prevent to cost-efficiently solve the voltage control problem. As an alternative, distributed strategies, which require communication capabilities between neighbouring agents, are also considered. Such approaches enable resources (that are physically close) to share information in order to cooperatively achieve the desired target levels, while considering other objectives such as losses minimization [13–15]. However, to further improve the optimality of the control solution, centralized voltage control algorithms, which are mainly based on an Optimal Power Flow (OPF) formulation, have also been proposed [16–18].

In general, although the latter centralized model-based techniques have shown promising performance, they are plagued with two main issues.

Firstly, they require to solve challenging optimization problems, which are non-linear and non-convex (from the AC power flow equations used to comply with the physical constraints of the electrical distribution system), and subject to uncertainties (from the stochastic load and generation changes, and the unexpected contingencies). The OPF-based methods thereby face scalability issues, which makes them of little relevance for real-time operation. This is partly addressed by using efficient nonlinear programming techniques [19], or through convex approximations of power flow constraints, which mainly resort on second order cone programs [20] or linear reformulations using the sensitivity analysis [21–23]. However, modeling errors inevitably arise and may lead to unsafe and sub-optimal solutions. Moreover, the recent trend of operating the modern distribution networks in closed loop mode makes traditional approximations even less accurate [24].

Secondly, the common feature of model-based techniques is that they assume that the physical parameters of the distribution networks are perfectly known, which is impractical due to the high complexity of these systems. In that regard, the real-time characteristics of the network components are not static, and are governed by complex dynamic dependencies [25]. For instance, deviations of parameters can arise from the atmospheric conditions and aging. Important effects are thereby often neglected, i.e., load power factors are not available precisely, there is a complex dependence structure between load and voltage levels, line impedances vary with the conductor temperatures, and the shunt admittances of lines as well as the internal resistance of transformers are also affecting network conditions [26].

The first issue (related to the high computational costs of model-based control algorithms) has led to the implementation of reinforcement learning (RL). These data-driven methods have the advantage to directly learn their operating strategy from historical data in a model-free fashion (without any assumptions on the functional form of the model). Consequently, they can show good robustness under very complex environments with measurement noise [27,28]. A novel deep reinforcement learning (DRL)-based voltage control scheme (named Grid Mind) is developed in [29]. In particular, two different techniques have been compared, i.e., deep Q-network (DQN) and deep deterministic policy gradient (DDPG), and both have shown promising outcomes. In [30], voltage regulation is improved using a RL-based policy that determines the optimal tap setting of transformers. Then, a new voltage control solution combining actions on two different time scales is implemented in [31], where DQN is applied for the (slow) operation of capacitor banks. Finally, multi-agent frameworks have been developed in [32,33] to enable decentralized executions of the control procedure that do not require a central controller.

However, all these methods are disregarding the endogenous uncertainties on network parameters, which may mislead the DSO into believing that the control strategy satisfies technical constraints, while it may actually result into unsafe conditions. In this context, the main contribution of this paper is to propose a self-learning voltage control tool based on deep reinforcement learning (DRL), which accounts for the limited knowledge on both the network parameters and the future (very-short-term) working conditions. The proposed tool can support DSOs in making autonomous and quick control actions to maintain nodal voltages within their safe range, in a cost-optimal manner (through the optimal use of ancillary services in a market environment). In this work, it is assumed that for voltage control purpose, we can act on the active and reactive powers of DG units as well as on the transformer tap position. The resulting problem is formulated as a single-agent centralized control model.

The main advantage of the proposed method lies in its ability to learn from scratch (in an off-line fashion) and gradually master the system operation. Hence, the computational burden is transferred in pre-processing (when the model is calibrated/learned through many simulations), such that the real-time control process (in actual field operation when the agent is trained) is insignificant ($\ll$1 s). Also, the model-free tool allows to immunize the voltage control procedure against uncertainties in both exogenous (load conditions) and endogenous (network parameters) variables, while accounting for approximations in the power flow models describing the system operation. Results from a case study on a 77-bus, 11 kV radial distribution system reveal that the proposed tool allows determining an optimal policy that lead to safe grid operation at low costs.

The remainder of this paper is organized as follows. Section 2 introduces the theoretical background in reinforcement learning, with a particular interest on the deep deterministic policy gradient (DDPG) algorithm, which allows to handle high-dimensional (and continuous) action spaces. Section 3 describes the simulation environment, including the different sources of uncertainty. The developed method is tested (using new representative network conditions) in Section 4 on a realistic 77-bus system, where we validate its robustness through the numerical simulations. Finally, conclusions and perspectives for future research are given in Section 5.

2. Reinforcement Learning Background

In this section, we introduce the basics of reinforcement learning (RL), while making the practical connections with the voltage control problem.

2.1. Markov Decision Process

Firstly, the problem has to be formulated as a Markov Decision Process (MDP). The general principle consists of an agent interacting with an environment $\mathcal{E}$ over a number of discrete time steps until the agent reaches a terminal state. In particular, at each step t, the agent observes a state s_t from the state space $\mathcal{S}$, and selects an action $a_t \in \mathcal{A}$ according to its policy $\pi(a_t|s_t)$. As a result, the agent ends up in the next state $s_{t+1} \sim \mathcal{P}(s_{t+1}|s_t, a_t)$ while receiving an immediate scalar reward r_t based on the distribution $\mathcal{R}(r_t|s_t, a_t)$ in accordance with the natural laws of the environment. The next state s_{t+1} depends only on the action a_t on state s_t (and not on the prior history), which is a characteristic referred to as the Markov property.

In this work, the agent is the central controller which regulates the voltage level within its control area, and the environment is the electrical distribution network (including the realization of the different sources of uncertainty affecting its operation). The state-transition model $\mathcal{P}(s_{t+1}|s_t, a_t)$ and the reward function $\mathcal{R}(r_t|s_t, a_t)$ are inherently stochastic, and the problem can thus be formulated using reinforcement learning.

2.1.1. State Space

The state space of the RL agent (i.e., central controller) is defined by the information that can be measured in real-time by SCADA (Supervisory Control and Data Acquisition) or PMU

(Phasor Measurement Unit). In that regard, the state space s_t at time t contains the voltage levels $V_{n,t}$ for each node $n \in \mathcal{N}$ of the distribution system. Then, this information is complemented by the (predicted) maximum power level $\overline{P}_{g,t+1}$ of each distributed generators $g \in \mathcal{G}$ at the next time interval $t+1$. This information (reflecting, e.g., the maximum energy contained in the wind) allows the agent to know the upper limit of these control variables when taking its actions. Practically, this is achieved using a (deterministic) single-step ahead forecaster, which is based on an advanced architecture of recurrent neural networks, as presented in [34]. The latter is tailored to predict the power level in the upcoming future by leveraging the past dynamics of the generator output. Finally, the current tap position Tap_t of the transformer (which defines the turn ratio between primary and secondary voltage levels) is also included in the state space s_t.

$$s_t = (V_{1,t}, ..., V_{N,t}, \overline{P}_{1,t+1}, ..., \overline{P}_{G,t+1}, Tap_t) \tag{1}$$

2.1.2. Action Space

The action space a_t to fix voltage issues in the studied network consists in changing the active and reactive powers of DG units, i.e., $\Delta P_{g,t}$ and $\Delta Q_{g,t}$ $\forall g \in \mathcal{G}$, as well as adjusting the transformer tap ratio ΔTap_t.

$$a_t = (\Delta P_{1,t}, ..., \Delta P_{G,t}, \Delta Q_{1,t}, ..., \Delta Q_{G,t}, \Delta Tap_t) \tag{2}$$

The actual changes in active $\Delta P_{g,t}$ and reactive $\Delta Q_{g,t}$ power levels initiated at time t (which will define the power output at time $t+1$) are limited by the available power at time $t+1$:

$$0 \leq P_{g,t} + \Delta P_{g,t} \leq \overline{P}_{g,t+1} \; \forall g \in \mathcal{G} \tag{3}$$

$$\underline{\Delta Q}_{g,t+1} \leq \Delta Q_{g,t} \leq \overline{\Delta Q}_{g,t+1} \; \forall g \in \mathcal{G} \tag{4}$$

where $P_{g,t}$ denotes the power level of (dispatchable) generator g at time t, while $\underline{\Delta Q}_{g,t+1}$ and $\overline{\Delta Q}_{g,t+1}$ determine the safe range of variation of reactive power of unit g around the operation point. Likewise, the variation ΔTap_t around the operation point Tap_t of the transformer tap change is given by:

$$\underline{\Delta Tap} \leq Tap_t + \Delta Tap_t \leq \overline{\Delta Tap} \tag{5}$$

where $\underline{\Delta Tap}$ and $\overline{\Delta Tap}$ are the physical limits of the on-load tap changer.

It should be noted that other types of control actions, such as changing the terminal voltage set-points of (medium-sized) conventional generators or switching shunt devices, could also be considered if such resources are available in the system.

2.1.3. Reward

As the goal of the algorithm is to eliminate voltage issues at a minimal cost, the reward includes both the costs inherent to control actions (changing set-points of control variables, which may reflect the costs of relying on ancillary services [35]), and the costs of violating network constraints (which may damage the equipment). In this way, the immediate reward r_t at time step t is defined as follows:

$$r_t = -\sum_{g \in \mathcal{G}} \left(C_Q |\Delta Q_{g,t}| + C_P |\Delta P_{g,t}| \right) - C_{TR} |\Delta Tap_t| + \begin{cases} +R_{pos}, \; \forall \, V_{n,t} \in [\underline{V}, \overline{V}] \\ -R_{neg}(\underline{V} - V_{n,t}), \; \forall \, V_{n,t} < \underline{V} \\ -R_{neg}(V_{n,t} - \overline{V}), \; \forall \, V_{n,t} > \overline{V} \end{cases} \tag{6}$$

where $\underline{V}$ and $\overline{V}$ are respectively the lower and upper bounds delimiting the safe voltage levels. Then, coefficients C_P and C_Q represent the costs of modifying the active and reactive powers of DG units, while C_{TR} stands for the (high) cost of changing the transformer tap position. Typically, we have $C_Q < C_P < C_{TR}$. Indeed, modifying the reactive power of generators can be done at almost no cost (using the power electronics converters), while the curtailment of active power infers a loss of generated energy that ultimately results into a financial loss [36]. Then, high costs are associated with a tap change of the transformer due to the aging effects on the tap changer contacts. The terms R_{pos} and R_{neg} respectively reflect the positive reward for the nodes having voltages within the safe range, and the negative reward (i.e., penalty) for nodes outside the permitted zone. In general, all these costs need to be properly weighted (see Section 4). Indeed, if the costs of actions C_Q, C_P and C_{TR} are too high with respect to R_{pos} and R_{neg}, the agent may choose to suffer the negative rewards related to voltage violations (rather than correcting the voltage problem). Conversely, if the costs of actions are too low, unnecessary actions may be taken (to ensure the positive rewards related to safe voltage levels).

2.2. Reinforcement Learning Algorithm

Like most machine learning techniques, it is important to differentiate training and test stages.

During the training, the goal of the agent is to learn the best policy π^*, i.e., to select actions that maximize the cumulative future reward $G_t = \sum_{j=t}^{T} \gamma^{j-t} r_j$ with a discount factor $\gamma \in [0,1]$. This can be achieved by approximating the optimal action-value function $Q^*(s,a) = \mathbb{E}^{\pi^*}(G_t|s,a)$, which is the expected discounted return of taking action a in state s, then continuing by choosing actions optimally. Indeed, once Q^*-values are obtained, the optimal policy can be easily constructed by taking the action given by $a_t^* = \text{argmax}_{a \in \mathcal{A}} Q^*(s_t, a)$. Using Bellman's principle of optimality, $Q^*(s_t, a_t)$ can be expressed as

$$Q^*(s_t, a_t) = \mathbb{E}_{s_{t+1} \sim \mathcal{E}} \left[r_t + \gamma \max_{a_{t+1}} Q^*(s_{t+1}, a_{t+1}) \right] \tag{7}$$

where the next state s_{t+1} is sampled from the environment's transition rules $\mathcal{P}(s_{t+1}|s_t, a_t)$. In general, an agent starts from an initial (poor) policy that is progressively improved through many experiences (during which the agent learns how to maximize its rewards).

When the training is completed, i.e., during the test (in practical field operations), the trained agent selects the greedy action a_t^* according to its learned policy.

This general principle is the source of many different RL algorithms, each with different characteristics that suits different needs. In this context, the choice of the most suited technique for the voltage control task is mainly driven by the fact that both state and action spaces are continuous. Hence, well-known algorithms, such as (deep) Q-learning, will not be considered as they only deal with a discrete action space. In this work, we will thereby focus on the deep deterministic policy gradient (DDPG) technique.

2.3. Deep Deterministic Policy Gradient (Ddpg) Algorithm

The deep deterministic policy gradient (DDPG) relies on a complicated architecture, referred to as actor-critic [37], which is depicted in Figure 1. The goal of the actor is to learn a deterministic policy $\mu_\phi(s)$ which selects the action a based on the state s. The quality of the action is estimated by the critic, by computing the corresponding $Q_\theta(s,a)$. To achieve good generalization capabilities of both actor and critic functions, they are estimated using deep neural networks, which are universal non-linear approximators that are very robust when the state and action spaces become large.

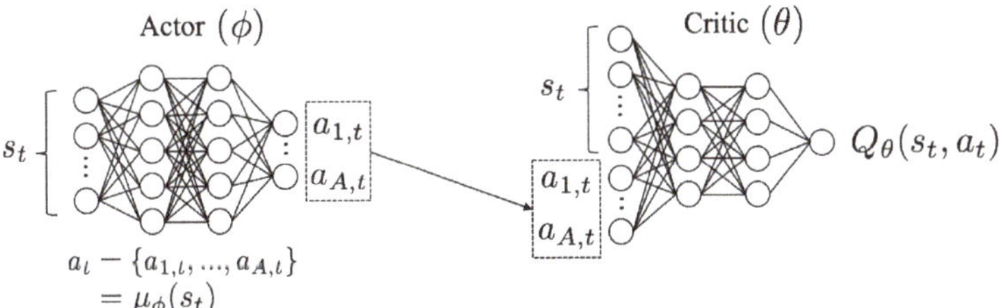

Figure 1. Working principle of the DDPG agent, which relies on an actor-critic architecture.

Overall, starting from an initial state s, the actor neural network (characterized by weight parameters ϕ) determines the action $a_t = \mu_\phi(s_t)$. This action is then applied to the environment, which yields the reward r_t and the next state s_{t+1}. The experience tuple (s_t, a_t, r_t, s_{t+1}) is then stored in the replay memory. Once the replay memory includes enough experiences, a random mini-batch of D experiences is sampled. For each sample in the mini-batch, the state s_t and the action a_t are fed into the critic neural network (characterized by weight parameters θ) that yields the Q-value. Both networks are then jointly updated, and the procedure is iterated until convergence.

Practically, the critic network is trained by adjusting its parameters θ_i (at regular intervals $i \in \mathcal{I}$ during the learning phase) so as to minimize the mean-squared Bellman error (MSBE) (9). In contrast to supervised learning, the actual (i.e., optimal) target value $r_t + \gamma \max_{a_{t+1}} Q^*(s_{t+1}, a_{t+1})$ is unknown, and is thus substituted with an approximate target value y_t (using the estimation Q_{θ_i}):

$$y_t = r_t + \gamma \max_{a_{t+1}} Q_{\theta_i}(s_{t+1}, a_{t+1}) \tag{8}$$

where a_{t+1} is given by the critic network, i.e., $a_{t+1} = \mu_{\phi_i}(s_{t+1})$.

Contrary to supervised learning where the output of the neural network and the target value (i.e., ground truth) are completely independent, we see that the target value y in (8) depends on the parameters θ_i and ϕ_i that we are optimizing in the training. This link between the critic's output $Q_{\theta_i}(s_t, a_t)$ and its target $r_t + \gamma \max_{a_{t+1}} Q_{\theta_i}(s_{t+1}, a_{t+1})$ may infer divergence in the learning procedure. A solution to this problem is to use separate target networks (for both the critic and the actor), which are responsible for calculating the target values. Practically, these target networks are time-delayed copies of the original networks with parameters $\theta_{i,\text{targ}}$ and $\phi_{i,\text{targ}}$ that slowly track the (reference) learned networks. As explained in [37], these target networks are not trained, and enable to break the dependency between the values computed by the networks and their targeted value, thereby improving stability in learning.

As a result, the critic network is trained (i.e., updated) by minimizing the following MSBE loss function $\mathcal{L}(\theta_i)$ with stochastic gradient descent:

$$\mathcal{L}(\theta_i) = \sum_D \left(\underbrace{Q_{\theta_i}(s_t, a_t)}_{(i)} - \underbrace{\left(r_t + \gamma Q_{\theta_{i,\text{targ}}}(s_{t+1}, \mu_{\phi_{i,\text{targ}}}(s_{t+1})) \right)}_{(ii)} \right)^2 \tag{9}$$

Starting from random values $\theta_{i=0}$, the parameters θ_i are thus progressively updated towards the optimal action-value function Q^* by minimizing the difference between (i) the output of the critic and (ii) the target (computed with target networks), which provides an estimate of the Q-function using both the outcome r_t of the simulation model and the action a_{t+1} from the target actor network. The update is performed on a mini-batch D of different experiences $(s_t, a_t, r_t, s_{t+1}) \sim U(D)$, drawn uniformly at random from the pool of historical samples. This (replay buffer) procedure breaks the similarity between consecutive training samples, thus avoiding that the model is updated towards a local minimum.

In parallel, the actor network is trained (on the same mini-batch D) with the goal of adapting its parameters ϕ_i, so as to provide actions a_t that maximize Q_{θ_i}. This amounts to maximize the following function $\mathcal{L}(\phi_i)$, which is achieved with a gradient ascent algorithm:

$$\mathcal{L}(\phi_i) = \sum_D Q_{\theta_i}\left(s_t, \mu_{\phi_i}(s_t)\right) \tag{10}$$

To ensure that the DDPG algorithm properly explores its environment during the training phase, noise ϵ_t is added to the action space, i.e., $a_t = \mu_\phi(s_t) + \epsilon_t$. In particular, we use an exponential decaying noise so as to favor exploration at the start of the training, which is then progressively decreased to stimulate exploitation as the agent converges towards the optimal policy. Naturally, when the model is trained (and used during test time), no noise is added to the optimal action a^*.

3. Simulation Environment

To train the DRL agent, it is necessary to build a simulation environment $\mathcal{E}$ that mimics the actual system. This environment is composed of three modules: (i) to generate realistic deviations of the expected nodal load and distributed generation powers for the next time step (to reflect prediction errors), (ii) to provide realistic values of the uncertain network parameters, and (iii) to simulate the physical flows in the distribution network.

As depicted in Figure 2, the RL agent is trained off-line through interactions with the simulation environment, which allows calibrating the RL model using experience and rewards. As previously explained, the starting point is an observation of the state s_t of the environment (e.g., nodal voltage levels of the distribution system). Based on this information, the (target) actor network is used to take an action $a_t = \mu_\phi(s_t) + \epsilon_t$ (where the additional noise ϵ_t is used during training to boost exploration). It should be noted that, if no voltage problem is observed, the optimal action is to do nothing. Then, the simulator (thoroughly described in the rest of this Section) is used to determine the impact of the action on the environment, which consists in computing the reward r_t, but also the next state s_{t+1}. Then, the (target) critic is used to evaluate the quality of the decisions, and both actor and critic networks are updated using respectively (10) and (9) to improve the policy of the DRL-based agent.

When the learning is performed, the agent can be deployed for practical power system operation (for which only the actor network is useful). Interestingly, the agent can still continue its learning (and thus adapt to potential misrepresentations of the simulation environment) by adjusting its parameters through on-line feedback. This may also serve for calibrating the model to the time-varying conditions of the system.

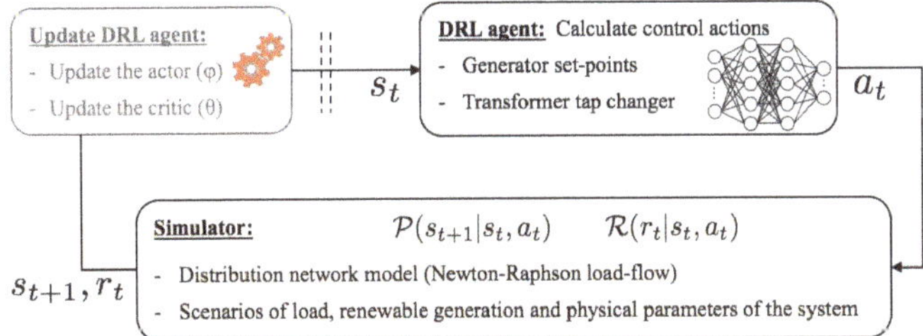

Figure 2. Training of the DRL agent for autonomous voltage control in distribution systems.

3.1. Exogenous Uncertainties on the Network Operating Point

The first category of uncertainties belongs to the network working point (regarding both the nodal consumptions and generations). Indeed, the output power of renewable-based generators is intermittent upon the nature of their primal sources (mainly wind and solar), such that the generated power can quickly vary within a short interval. Moreover, the nodal consumption and generation levels are not always measurable. Consequently, in practice, the future operating state of the distribution system is not known with certainty, and this stochasticity is here represented with scenarios of representative prediction errors. Practically, for the renewable generation, a database is constructed based on the historical prediction errors of the employed forecaster (described in Section 2.1.1), and a sample is randomly drawn from this database to generate the desired scenario. For the nodal loads, the same sampling strategy is used to simulate the (uncertain) changes in the consumption level.

3.2. Endogenous Uncertainties on the Network Component Models and Parameters

The second category of uncertainties is related to partial knowledge of network component models and parameters. In general, network analyses and simulations are carried out relying on the simplified models of network components, which do not correctly represent the physical relations and dependencies within the real network. This includes uncertainties associated with the line, load and transformer models [26].

In particular, we model the thermal dependency phenomenon whereby the line resistance fluctuates with respect to the conductor temperature variation. Then, the uncertainty associated with the load power factor is considered to better reflect the different natures, types and amplitudes of the various load demands. Moreover, as shown in [38], the internal resistance of the transformer can have a significant effect on the node voltages, and is thereby also incorporated in the network model. Finally, in contrast to typical network models, the shunt admittances of power lines are taken into account using the PI line model. Overall, all these (uncertain) parameters are modelled as random variables changing within representative predefined bounds.

3.3. Distribution Network Model

The electrical network operation is modeled through load-flow calculations, which are solved using the Newton-Raphson approach.

4. Case Study

To solve the voltage control problem, the DDPG algorithm is implemented in Python using PyTorch and Gym libraries. The solution is tested on the 11 kV radial distribution system with $N = 77$ buses shown in Figure 3 [39]. The bus 1 is the high-voltage (HV) connection point, which is considered as the slack node. The substation (between nodes 1 and 2) supplies 8 different feeders, for a total of 75 loads. The maximum (peak) active and reactive consumption powers equal to 24.27 MW and 4.85 Mvar, respectively. The system is also hosting 22 (identical) distributed generators, with an installed power equal to 4 MW.

The objective of the DRL-based agent is to maintain the voltage magnitudes of the 77 buses within the desired range. In order to illustrate the effectiveness of the proposed control scheme, these allowed voltage limits are defined by a very conservative range of [0.99, 1.01] p.u., and the initial reactive powers of DGs are set to zero. The reward function (6) is characterized by a compromise between the costs of voltage violations and those of corrective actions. We give more weight in maintaining safe voltage levels by defining $R_{pos} = 0.1$ and $R_{neg} = 15$, while C_{TR}, C_P and C_Q are respectively set to 1, 0.1 and 0.04.

A total of 12,000 initial operating states (that need to be processed by the DRL-based agent) are generated with the simulation model, among which 10,000 are used to train the agent, while the remaining 2000 scenarios are kept (as a test set) to evaluate the performance of the resulting model. It should be noted that, in this work, the agent has a single step to process each of the generated scenarios (it cannot rely on several interactions with the environment to solve a voltage problem). The value of the discount factor γ is thereby fixed to 1.

To have an overview of the global network conditions in the case where no control action is performed, we show in Figure 4 the distribution of nodal voltage levels (for the 12,000 simulated states) using a boxplot representation. We observe that violations of voltage limits [0.99, 1.01] p.u. occur more than 50% of the time. In particular, the distribution is asymmetrical, skewed towards more over-voltage issues (due to the high penetration of distributed generation) which occurs in 40.1% of the simulated samples.

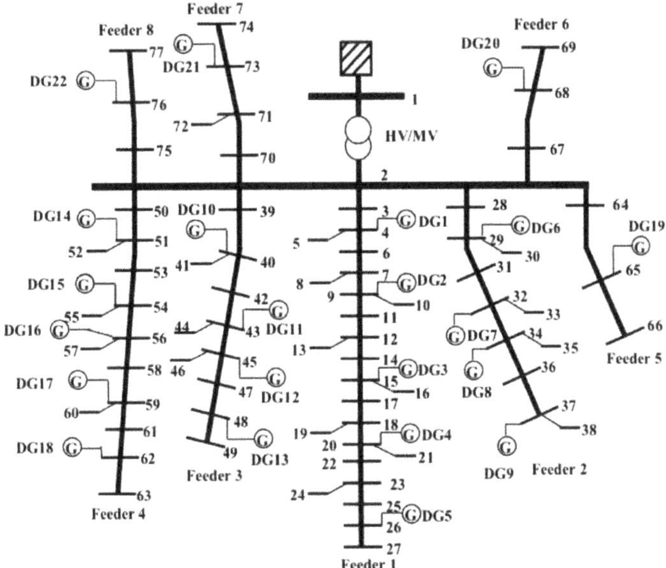

Figure 3. Schematic diagram of the 77-bus distribution system. The section between bus 1 and 2 is the substation, which is supplying 8 different feeders.

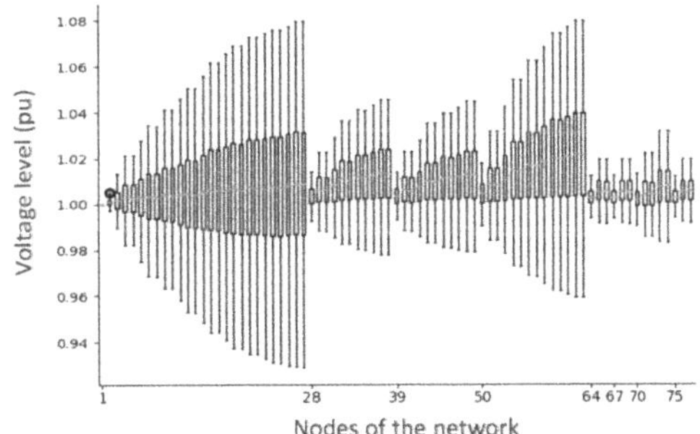

Figure 4. Boxplot representing the (nodal) distributions of the voltage levels for the 77 buses among the 12,000 simulated states.

4.1. *Impact of Ddpg Parameters*

In the proposed case study, the state space s_t is of size 100, i.e., 77 dimensions for the nodal voltages $V_{n,t}$, 22 dimensions for the (predicted) maximum power of the 22 generators $\overline{P}_{g,t+1}$, and 1 dimension for the position of the tap changer Tap_t. Also, the action space a_t is of size 45, i.e., $2 \times 22 = 44$ dimensions corresponding to the changes in active and reactive power for the 22 generators, and 1 dimension for changing the position of the tap changer. Hence, as sketched in Figure 5, the actor network has an input layer of size 100 (i.e., composed of 100 neurons), and an output layer of size 45. Then, the critic network is characterized by 145-dimensional input layer, for a single output.

Based on this (fixed) information, we then performed an optimization of the hyper-parameters of the DRL-based agent, which consists in optimizing its complexity by adding extra hidden layers in the architecture of both actor and critic neural networks. In particular, the best performance was achieved by connecting the input and output layers (for both the actor and the critic networks) with 5 fully connected layers, with 20 units in all layers. The activation functions of the hidden layers are ReLU (rectified linear units). Then, the hyperbolic tangent function is used for the output layer of the actor, while a linear function is employed for the critic. The batch size of the learning is set to 16 samples, and the target networks are updated (during the training) with a delay of 10 iterations. Both actor and critic networks are initialized with random weights in the range $[-0.1, 0.1]$.

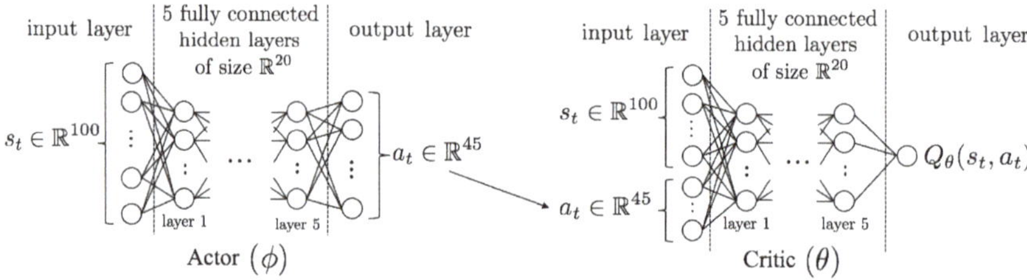

Figure 5. Representation of the neural network architectures for both actor and critic.

The exploration-exploitation parameter (i.e., extra noise added to the actions during the training) is $\epsilon_t = \mathcal{N}(0, 0.2) \times (0.005 + 0.995e^{-k/\Delta_T})$, where $\mathcal{N}(0, 0.2)$ is a zero-mean Gaussian noise with a standard deviation of 0.2, which is exponentially decaying along the training iterations k. The decay period Δ_T is equal to 5000 episodes. In general, this action noise has a significant impact on the learning abilities of the DRL-based agent. This observation is illustrated in Figure 6, where we depict two different learning curves where all parameters of the agents are similar, except for the action noise. In particular, the optimal calibration of $\mathcal{N}(0, 0.2)$ is compared to a perturbation of $\mathcal{N}(0, 0.6)$ (with the same decaying intensity over the training samples).

In general, when the perturbations are too small, the training may fail to properly explore the search space (which increases the probability to end up in a local minimum), while oversized perturbations may negatively affect the learning (and even leading the algorithm to repeatedly perform the same action).

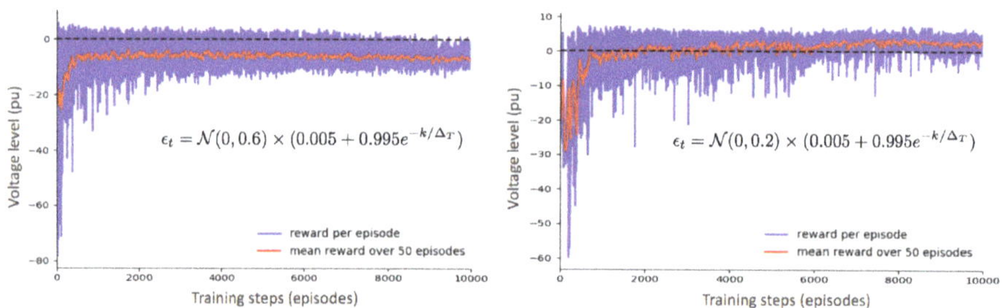

Figure 6. Evolution of the total immediate rewards r_t across training episodes for two different configurations of the action noise ϵ_t.

For the best model (right part of Figure 6), we see that the DDPG control scheme quickly learns (after around 7500 interactions with the environment) a stable and efficient policy. In particular, at the beginning (during the 2000 first training steps), the agent randomly selects actions, which lead to many situations where it deteriorates the electrical network conditions. However, in the course of the learning procedure, the agent is progressively evolving, and starts solving the voltage issues with less costly decisions. The agent eventually converges to total rewards $r \approx 5$. In contrast, the other model (left part of Figure 6) achieves convergence at a much lower performance (total rewards of $r \approx -7.5$), which roughly corresponds to the same reward as when no action is performed. In general, the main advantage of the proposed framework lies in its generic design that makes it broadly applicable (e.g., to any distribution system), and in its ability to adapt to the varying operating conditions. Evidently, when the methodology is applied to another environment, the DDPG agent needs to be re-trained from scratch, and its hyper-parameters (e.g., training noise, as well as number of hidden layers and number of neurons for both actor and critic networks) also need to be adapted.

4.2. Impact of Endogenous Uncertainties

The impact of endogenous uncertainties (regarding the physical parameters of the distribution system) is evaluated through the analysis of three cases.

1. The network parameters are considered as perfectly known in both training and test stages;
2. The uncertainty on the network parameters are neglected during the training phase (to mimic current optimization models), but are considered when evaluating the performance of the trained DRL-based agent (to reflect reality);
3. The uncertainty on the values of network parameters is accounted for in both training and test phases.

The simulation results regarding the three cases are summarized in Figure 7. Practically, we represent the evolution of the negative reward (which is a measure of the voltage violations) in both training and test phases. This negative reward r_{neg} is equal to 0 in the perfect situation where all nodal voltages pertain to $[\underline{V}, \overline{V}]$ = [0.99, 1.01] p.u., and decreases in negative values with the severity of voltage violations, i.e.,:

$$r_{\text{neg}} = \begin{cases} 0, \ \forall V_n \in [\underline{V}, \overline{V}] \\ -R_{neg}(\underline{V} - V_n), \ \forall V_n < \underline{V} \\ -R_{neg}(V_n - \overline{V}), \ \forall V_n > \overline{V} \end{cases} \tag{11}$$

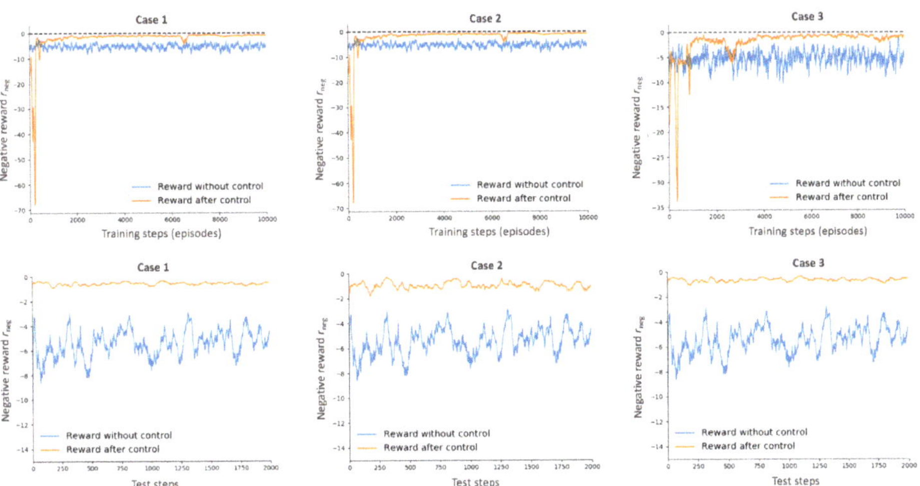

Figure 7. Evolution of the reward r_{neg} in the three studied cases in both training and test stages.

We observe that when uncertainties associated with the model parameters are neglected during the training (cases I and II), the RL agent quickly find actions that remove voltage violations, i.e., the upper bound of the negative reward $r_{\text{neg}} = 0$ is almost reached in around 2000 episodes. This performance is achieved in more than 4000 episodes when dealing with endogenous uncertainties due to the increased difficulty of the task. This effect is also translated into a higher variability of the reward. Interestingly, by comparing the evolution of r_{neg} with the total reward r in Figure 6 during the training, we see that even though the agent is able to mitigate the voltage issues after 4000 training episodes, the cost-efficiency of the actions can still be improved (which is realized during the next 4000 episodes).

To quantify the impact of neglecting the endogenous model uncertainties, the mean value of the negative reward r_{neg} in (11) over the last 2000 episodes of the training phase, and over the 2000 new episodes of the test set are provided in Table 1 for the three studied cases.

Table 1. Average value of the negative reward r_{neg} across training and test sets.

	Case 1	Case 2	Case 3
Training set	−0.6	−0.6	−0.69
Test set	−0.53	−0.93	−0.67

As expected, the agent that is agnostic to endogenous uncertainties on the physical parameters of the system during the training (cases 1 and 2) achieves a lower out-of-sample performance when these effects are modeled in the test set. Specifically, the reward r_{neg} drops from −0.53 (in case 1 when endogenous uncertainties are also disregarded at the test stage) down to −0.93 in the realistic case 2. In this latter situation, the agent expects a reward of around −0.6 (at the end of its learning), while it actually results in a disappointing ex-post outcome of −0.93. This problem can be efficiently alleviated by incorporating these endogenous uncertainties within the learning procedure. In that framework (case 3), the training and test rewards are close to each other, i.e., $r_{neg} \approx -0.67$, which illustrates the good performance of the proposed method.

4.3. Extreme Cases

In this part, the outcome of the DRL-based agent is illustrated for two extreme situations, respectively corresponding to the worst-case over- and under-voltage states. These states result from the combination of extreme consumption and generation conditions, associated with unfavourable parameters of the distribution system (such as high line impedances arising from a temperature increase).

In Figure 8, we select the scenario (from the 2000 test samples) which leads to the worst-case voltage rise. In this case, the load demands are low (globally equal to around 10% of their nominal values) while active powers of DGs are at 90% of their rated values. The initial system voltages significantly exceed the upper limit of 1.01 p.u. (for almost all nodes), and reach a maximum value of 1.08 p.u. at node 27 (end of feeder 1). Also, the absolute value of the reward associated with the control actions taken by the proposed DDPG algorithm is represented in the right part of the Figure 8.

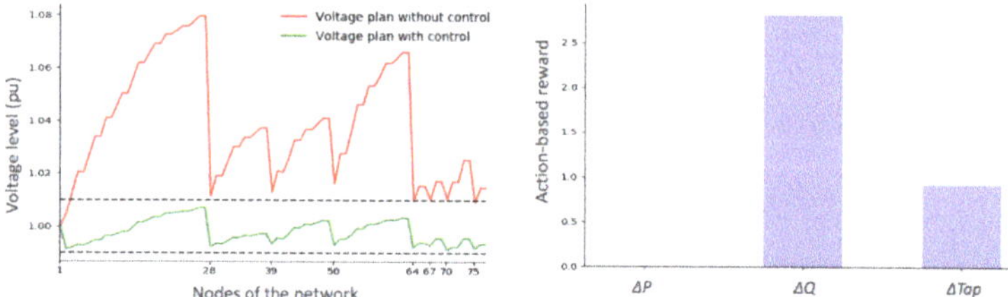

Figure 8. Initial nodal voltages as well as the corrected ones obtained by the DRL-based agent in an extreme over-voltage situation. The corresponding (absolute value) of the reward related to each family of actions is also displayed.

Interestingly, the DRL-based agent has completely solved the voltage problem. We see that it did not rely on the curtailment of the active power of distributed generators. Indeed, this solution is more expensive than consuming reactive power (which is thereby the privileged action). However, the transformer tap ratio had also to be modified (i.e., voltage drop between nodes 1 and 2) to prevent over-voltages at the end of the feeders.

In Figure 9, the voltage drop condition is analyzed, which corresponds to a situation where load demands are maximum, while active powers of DGs are equal to zero. This results into under-voltage issues in many nodes of the distribution system.

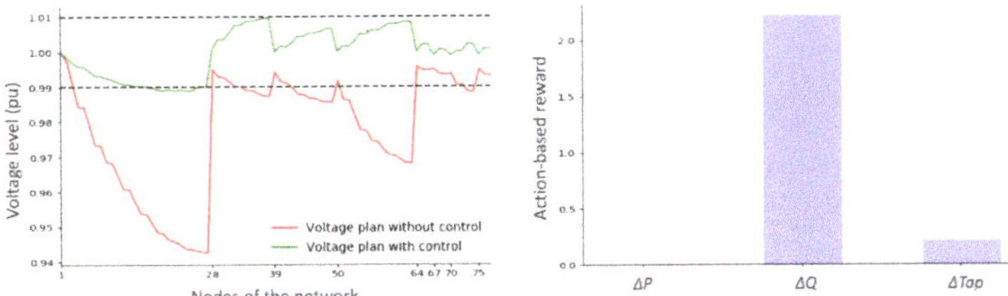

Figure 9. Initial nodal voltages as well as the corrected ones obtained by the DRL-based agent in an extreme under-voltage situation. The corresponding (absolute value) of the reward related to each family of actions is also displayed.

Similarly to the over-voltage case, the privileged action is to modify the reactive power level of DG units (here by exchanging capacitive reactive power to compensate the voltage drops). The corrected situation brings the voltage plan within the desired limits, at the exception of some nodes at the end of feeder 1 that are slightly violating the lower bound (of 0.99 p.u.).

In general, after the training, the agent is able to successfully make the right decisions. In particular, during the testing under new randomly generated conditions, the proposed DRL-based algorithm achieves robust solutions (against the various sources of uncertainty) that mitigate severe voltage violations using cost-effective actions.

5. Conclusions and Perspectives

This paper was devoted to the voltage control problem in distribution systems, which is facing new challenges from growing dynamics and uncertainties. In particular, current strategies are hampered by the limited knowledge of the network parameters, which may prevent achieving the optimal cost-efficiency. This problem is formulated as a centralized control of resources using deep reinforcement learning, through an actor-critic architecture that enables to properly represent the continuous environment. This framework bypasses the need to represent analytically the electrical system, such that the impact of model accuracy is decoupled from the control performance.

The main advantage of the proposed model is to put the computational complexity on the pre-processing (in a fully data-driven framework), such that the model provides very fast decisions in test time. Interestingly, the developed regulation scheme is not only easy to implement, but also cost-efficient as we observe that the agent is able to automatically adapt its behavior to varying conditions.

The promising outcomes of the work pave the way towards more advanced strategies, such as the extension to a decentralized approach using a multi-agent formulation (that would prevent the single point of failure of the centralized framework). Similarly, extending the framework to partially observable networks (where the state of the system is not fully known [40]) also offers a valuable area of research for system operators.

Author Contributions: Conceptualization, J.-F.T. and F.V.; methodology, J.-F.T.; validation, J.-F.T., and B.B.Z.; writing–original draft preparation, J.-F.T. and B.B.Z.; writing–review and editing, M.H., Z.D.G. and F.V.; supervision, F.V.; project administration, F.V. All authors have read and agreed to the published version of the manuscript.

Funding: This research received no external funding.

Acknowledgments: J.-F. Toubeau is supported by FNRS (Belgian National Fund of Scientific Research).

Conflicts of Interest: The authors declare no conflict of interest.

References

1. Toubeau, J.-F.; Vallée, F.; Grève, Z.D.; Lobry, J. A new approach based on the experimental design method for the improvement of the operational efficiency in Medium Voltage distribution networks. *Int. J. Electr. Power Energy Syst.* **2015**, *66*, 116–124. [CrossRef]
2. Wei, B.; Deconinck, G. Distributed Optimization in Low Voltage Distribution Networks via Broadcast Signals. *Energies* **2020**, *13*, 43. [CrossRef]
3. Klonari, V.; Toubeau, J.-F.; Lobry, J.; Vallée, F. Photovoltaic integration in smart city power distribution: A probabilistic photovoltaic hosting capacity assessment based on smart metering data. In Proceedings of the 2016 5th International Conference on Smart Cities and Green ICT Systems (SMARTGREENS), Rome, Italy, 23–25 April 2016; pp. 1–13.
4. Liu, H.J.; Shi, W.; Zhu, H. Distributed voltage control in distribution networks: Online and robust implementations. *IEEE Trans. Smart Grid* **2018**, *9*, 6106–6117. [CrossRef]
5. Ou-Yang, J.-X.; Long, X.-X.; Du, X.; Diao, Y.-B.; Li, M.-Y. Voltage Control Method for Active Distribution Networks Based on Regional Power Coordination. *Energies* **2019**, *12*, 4364. [CrossRef]
6. Klonari, V.; Toubeau, J.-F.; Grève, Z.D.; Durieux, O.; Lobry, J.; Vallée, F. Probabilistic simulation framework for balanced and unbalanced low voltage networks. *Int. J. Electr. Power Energy Syst.* **2016**, *82*, 439–451. [CrossRef]
7. Sun, H.; Guo, Q.; Qi, J.; Ajjarapu, V.; Bravo, R.; Chow, J.; Li, Z.; Moghe, R.; Nasr-Azadani, E.; Tamrakar, U.; et al. Review of challenges and research opportunities for voltage control in smart grids. *IEEE Trans. Power Syst.* **2019**, *34*, 2790–2801. [CrossRef]
8. Xiao, C.; Sun, L.; Ding, M. Multiple Spatiotemporal Characteristics-Based Zonal Voltage Control for High Penetrated PVs in Active Distribution Networks. *Energies* **2020**, *13*, 249. [CrossRef]
9. Xiao, C.; Zhao, B.; Ding, M.; Li, Z.; Ge, X. Zonal Voltage Control Combined Day-Ahead Scheduling and Real-Time Control for Distribution Networks with High Proportion of PVs. *Energies* **2017**, *10*, 1464. [CrossRef]
10. Zad, B.B.; Lobry, J.; Vallée, F. A Centralized approach for voltage control of MV distribution systems using DGs power control and a direct sensitivity analysis method. In Proceedings of the IEEE International Energy Conference (ENERGYCON), Leuven, Belgium, 2–4 April 2016.
11. Calderaro, V.; Conio, G.; Galdi, V.; Massa, G.; Piccolo, A. Optimal Decentralized Voltage Control for Distribution Systems With Inverter-Based Distributed Generators. *IEEE Trans. Power Syst.* **2014**, *29*, 230–241. [CrossRef]
12. Nowak, S.; Wang, L.; Metcalfe, M.S. Two-level centralized and local voltage control in distribution systems mitigating effects of highly intermittent renewable generation. *Int. J. Electr. Power Energy Syst.* **2020**, *119*, 1–15. [CrossRef]
13. Brenna, M.; Berardinis, E.D.; DelliCarpini, L.; Foiadelli, F.; Paulon, P.; Petroni, P.; Sapienza, G.; Scrosati, G.; Zaninelli, D. Automatic distributed voltage control algorithm in smart grids applications. *IEEE Trans. Smart Grid* **2013**, *4*, 877–885. [CrossRef]
14. Almasalma, H.; Claeys, S.; Mikhaylov, K.; Haapola, J.; Pouttu, A.; Deconinck, G. Experimental Validation of Peer-to-Peer Distributed Voltage Control System. *Energies* **2018**, *11*, 1304. [CrossRef]
15. Vovos, P.N.; Kiprakis, A.E.; Wallace, A.R.; Harrison, G.P. Centralized and Distributed Voltage Control: Impact on Distributed Generation Penetration. *IEEE Trans. Power Syst.* **2007**, *22*, 476–483. [CrossRef]
16. Capitanescu, F.; Bilibin, I.; Ramos, E.R. A comprehensive centralized approach for voltage constraints management in active distribution grid. *IEEE Trans. Power Syst.* **2013**, *29*, 933–942. [CrossRef]
17. Robertson, J.G.; Harrison, G.P.; Wallace, A.R. OPF Techniques for Real-Time Active Management of Distribution Networks. *IEEE Trans. Power Syst.* **2017**, *32*, 3529–3537. [CrossRef]
18. Guo, Q.; Sun, H.; Zhang, M.; Tong, J.; Zhang, B.; Wang, B. Optimal voltage control of pjm smart transmission grid: Study, implementation, and evaluation. *IEEE Trans. Smart Grid* **2013**, *4*, 1665–1674.
19. Qin, N.; Bak, C.L.; Abildgaard, H.; Chen, Z. Multi-stage optimization-based automatic voltage control systems considering wind power forecasting errors. *IEEE Trans. Power Syst.* **2016**, *32*, 1073–1088. [CrossRef]
20. Taylor, J.A. *Convex Optimization of Power Systems*; Cambridge University Press: Cambridge, UK, 2015.
21. Borghetti, A.; Bosetti, M.; Grillo, S.; Massucco, S.; Nucci, C.A.; Paolone, M.; Silvestro, F. Short-term scheduling and control of active distribution systems with high penetration of renewable resources. *IEEE Syst. J.* **2010**, *4*, 313–322. [CrossRef]

22. Zad, B.B.; Hasanvand, H.; Lobry, J.; Vallée, F. Optimal reactive power control of DGs for voltage regulation of MV distribution systems using sensitivity analysis method and PSO algorithm. *Int. J. Electr. Power Energy Syst.* **2015**, *68*, 52–60.
23. Pilo, F.; Pisano, G.; Soma, G.G. Optimal coordination of energy resources with a two-stage online active management. *IEEE Trans. Ind. Electron.* **2011**, *58*, 4526–4537. [CrossRef]
24. Loos, M.; Werben, S.; Maun, J.C. Circulating currents in closed loop structure, a new problematic in distribution networks. In Proceedings of the 2012 IEEE Power and Energy Society General Meeting, San Diego, CA, USA, 22–26 July 2012; pp. 1–7.
25. Rousseaux, P.; Toubeau, J.; Grève, Z.D.; Vallée, F.; Glavic, M.; Cutsem, T.V. A new formulation of state estimation in distribution systems including demand and generation states. In Proceedings of the 2015 IEEE Eindhoven PowerTech, Eindhoven, The Netherlands, 29 June–2 July 2015; pp. 1–6.
26. Zad, B.B.; Toubeau, J.-F.; Lobry, J.; Vallée, F. Robust voltage control algorithm incorporating model uncertainty impacts. *IET Gener. Transm. Distrib.* **2019**, *13*, 3921–3931.
27. Vlachogiannis, J.G.; Hatziargyriou, N.D. Reinforcement learning for reactive power control. *IEEE Trans. Power Syst.* **2004**, *19*, 1317–1325. [CrossRef]
28. Glavic, M.; Fonteneau, R.; Ernst, D. Reinforcement learning for electric power system decision and control: Past considerations and perspectives. *IFAC-PapersOnLine* **2017**, *50*, 6918–6927. [CrossRef]
29. Duan, J.; Shi, D.; Diao, R.; Li, H.; Wang, Z.; Zhang, B.; Bian, D.; Yi, Z. Deep-reinforcement-learning-based autonomous voltage control for power grid operations. *IEEE Trans. Power Syst.* **2019**, *35*, 814–817. [CrossRef]
30. Xu, H.; Dominguez-Garcia, A.; Sauer, P.W. Optimal tap setting of voltage regulation transformers using batch reinforcement learning. *IEEE Trans. Power Syst.* **2019**, *35*, 1990–2001. [CrossRef]
31. Yang, Q.; Wang, G.; Sadeghi, A.; Giannakis, G.B.; Sun, J. Two- timescale voltage control in distribution grids using deep reinforcement learning. *IEEE Trans. Smart Grid* **2019**, *11*, 2313–2323. [CrossRef]
32. Xu, Y.; Zhang, W.; Liu, W.; Ferrese, F. Multiagent-based reinforcement learning for optimal reactive power dispatch. *IEEE Trans. Syst. Man Cybern. Part (Appl. Rev.)* **2012**, *42*, 1742–1751. [CrossRef]
33. Wang, S.; Duan, J.; Shi, D.; Xu, C.; Li, H.; Diao, R.; Wang, Z. A Data-driven Multi-agent Autonomous Voltage Control Framework Using Deep Reinforcement Learning. *IEEE Trans. Power Syst.* **2020**. [CrossRef]
34. Toubeau, J.-F.; Bottieau, J.; Vallée, F.; Grève, Z.D. Deep Learning-Based Multivariate Probabilistic Forecasting for Short-Term Scheduling in Power Markets. *IEEE Trans. Power Syst.* **2019**, *34*, 1203–1215. [CrossRef]
35. Toubeau, J.-F.; Grève, Z.D.; Vallée, F. Medium-Term Multimarket Optimization for Virtual Power Plants: A Stochastic-Based Decision Environment. *IEEE Trans. Power Syst.* **2018**, *33*, 1399–1410. [CrossRef]
36. Olivier, F.; Aristidou, P.; Ernst, D.; Cutsem, T.V. Active Management of Low-Voltage Networks for Mitigating Overvoltages Due to Photovoltaic Units. *IEEE Trans. Smart Grid* **2016**, *7*, 926–936. [CrossRef]
37. Lillicrap, T.P.; Hunt, J.J.; Pritzel, A.; Heess, N.; Erez, T.; Tassa, Y.; Silver, D.; Wierstra, D. Continuous control with deep reinforcement learning. *arXiv* **2015**, arXiv:1509.02971.
38. Zad, B.B.; Lobry, J.; Vallée, F. Impacts of the model uncertainty on the voltage regulation problem of Medium Voltage distribution systems. *IET Gener. Transm. Distrib.* **2018**, *12*, 2359–2368.
39. Valverde, G.; Cutsem, T.V. Model predictive control of voltages in active distribution networks. *IEEE Trans. Smart Grid* **2013**, *4*, 2152–2161. [CrossRef]
40. Toubeau, J.-F.; Hupez, M.; Klonari, V.; Grève, Z.D.; Vallée, F. Statistical Load and Generation Modelling for Long Term Studies of Low Voltage Networks in Presence of Sparse Smart Metering Data. In Proceedings of the 42nd Annual Conference of IEEE Industrial Electronics Society (IECON), Florence, Italy, 23–26 October 2016.

Article

Impact of Lossy Compression Techniques on the Impedance Determination

Maik Plenz [1,*], Marc Florian Meyer [1], Florian Grumm [1], Daniel Becker [1], Detlef Schulz [1] and Malcom McCulloch [2]

1 Department of Electrical Power Systems, Helmut Schmidt University Hamburg, 22043 Hamburg, Germany; marc.meyer@hsu-hh.de (M.F.M.); florian.grumm@hsu-hh.de (F.G.); daniel.becker@hsu-hh.de (D.B.); detlef.schulz@hsu-hh.de (D.S.)

2 Department of Engineering Science, Oxford University, Oxford OX1 2JD, UK; malcolm.mcculloch@eng.ox.ac.uk

* Correspondence: maik.plenz@hsu-hh.de

Received: 3 June 2020; Accepted: 14 July 2020; Published: 16 July 2020

Abstract: One of the essential parameters to measure the stability and power-quality of an energy grid is the network impedance. Including distinct resonances which may also vary over time due to changing load or generation conditions in a network, the frequency characteristic of the impedance is an import part to analyse. The determination and analysis of the impedance go hand in hand with a massive amount of data output. The reduction of this high-resolution voltage and current datasets, while maintaining the fidelity of important information, is the main focus of this paper. The presented approach takes measured impedance datasets and a set of lossy compression procedures, to monitor the performance success with known key metrics. Afterwards, it continually compares the results of various lossy compression techniques. The innovative contribution is the combination of new and existing procedures as well as metrics in one approach, to reduce the size of the impedance datasets for the first time. The approach needs to be efficient, suitable, and exact, otherwise the decompression results are useless.

Keywords: impedance determination; lossy compression algorithms; singular value decomposition; wavelet transformation

---

1. Introduction

The energy system transformation and smart grid applications require knowledge about detailed power and load profiles with sophisticated datasets on the one hand. On the other, an increasing number of power electronic converters (PECs) from renewable energies and smart loads are integrated into the electrical supply system to measure and analyse the power-quality, stability, and control design considerations. Since the first generation of grid-connected converters, the grid impedance has been an important part of the analysis of the stability of the whole energy system or detection of islanding grids [1,2]. The so-called PECs are usually self-controlled pulse width modulation (PWM) power converters that connect generators or loads to the 50 Hz power supply system. For the power- quality analysis or a filter design of the converters detailed knowledge is required of the frequency characteristics of the network impedance at a specific grid-connection point [3]. Along with the increasing number of PECs especially in low and medium voltage grids, the generated, and transferred amount of data rises massively. This opens up multiple situations for system optimization. The current operational conditions of the municipal utilities, grid owner, or system operator—low bandwidth and low computational power—and the development of impedance shaping intensifies problems in many cases [4,5]. To improve memory consumption, collection, and transmission efficiency, the reduction of high-resolution impedance datasets while maintaining the fidelity of relevant information presents one opportunity for system optimization.

The overwhelming majority of the studies that try to tackle this larger issue focus on how to generate, analyse, and shape the network impedance so as to make it usable with the grid [6–8]. As regards the topic of compression, some approaches have been investigated in the field of medical data science [9].

This paper addresses the question of how to compress voltage and current datasets of an impedance measurement device by using lossy compression approaches without any detrimental effect on the impedance results of the measurement. Since raw voltage and current datasets contain further information e.g., about voltage harmonics, the aim is to compress the raw data instead of the calculated impedance data. Therefore, the scope of the paper is the grid impedance and the compression compatibility. It is not important to compress the data set one-to-one or to create a method with the highest efficiency, depending on processing duration or error level. The idea is to generate an easy to handle, efficient, sufficiently selective, accurate, and usable approach that output a compressed impedance dataset without irrelevancies. Two other highly important questions are, one, how to transfer and store large amounts of data using small amounts of resources and costs and, two, how to extract necessary information from the dataset. Due to limited computational i.e., bandwidth and storage space, and human resources lossy compression algorithms are promising. The new procedure and model technique, which combines measured impedance datasets, lossy compression techniques, and key metrics addresses exactly this specific gap in knowledge.

Section 2 describes the background of the impedance measurement to show the used test case and the simulation approach, which produced the dataset. The remainder of this paper is organized as follows. Section 3 presents typical lossy compression approaches, which can be used to reduce the amount of data. After those techniques have been explained, the approach taken in this paper, and the key metrics are introduced in Section 4. The obtained performance results are presented in Section 5 to show if they meet the required criteria. Finally, conclusion and outlook are presented in Section 6.

2. Mid Voltage Impedance Measurement System

In the literature, the methods to measure the network impedance may be categorized into active and passive methods used for power systems during operation. Active methods use excitation signals at the point of common coupling (PCC) to identify the impedance. The signal generator can be a current or voltage source or a current sink.

2.1. Impedance Identification

The used signal generation is a sink: A load resistor is switched on and off in a random pattern. Hence, the load current is a random pulse pattern. Figure 1 shows the principle connection scheme of the network impedance measurement (NIM) device and its corresponding complex equivalent circuit. The voltage in off and on state ($v_1(t)$ and $v_2(t)$) as well as the current in the on state $i_2(t)$ (off state current $i_1(t)$ is zero) are measured. The frequency-dependent complex values $\underline{V}_1(\omega), \underline{V}_2(\omega)$ and $\underline{I}_2(\omega)$, derived from the fast Fourier transform, are used to calculate the complex frequency-dependent impedance $\underline{Z}_\mathrm{N}(\omega)$ (1).

$$\underline{Z}_\mathrm{N}(\omega) = \frac{\underline{V}_2(\omega) - \underline{V}_1(\omega)}{\underline{I}_2(\omega)} \tag{1}$$

2.2. Impedance Identification of Three Phase Systems

In this section, the measurement system for the determination of the frequency dependency of the grid impedance is presented. The device includes highly accurate sensors for the measurement of voltage and current wave forms. Figure 2 shows a simplified scheme of the measurement setup to evaluate the impedance of the three-phase mid voltage PCC [10,11]. A resistive load is switched by an insulated gate bipolar transistor (IGBT), while the measurement loop (e.g., L1–L2) is selected by a B6 thyristor bridge. A 3D model of the measurement device and its main components are depicted in

Figure 3. The measurement device is portable and may be connected to connection points in medium voltage grids via its own medium voltage switch gear (SF6 circuit breaker). To determine the line impedances of a three-phase system four measurements are required:

- At first, the open circuit is measured to obtain the reference $\underline{V}_1(\omega)$. Hence, the load is not pulsed and $\underline{I}_1(\omega)$ is zero,
- Then, a pulse pattern is applied to the three loops of phase a to b, phase b to c, and phase c to a.

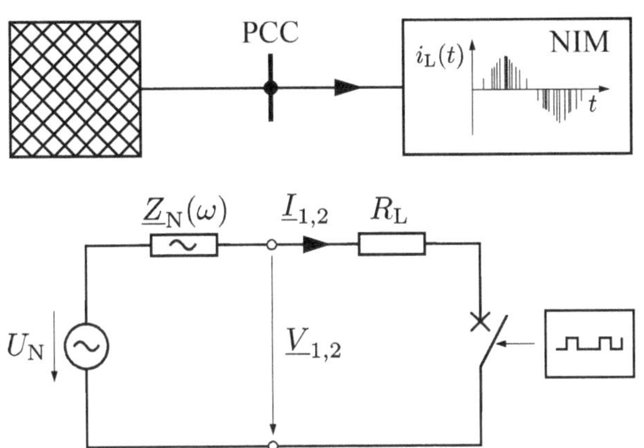

Figure 1. Principle of the network impedance measurement/identification.

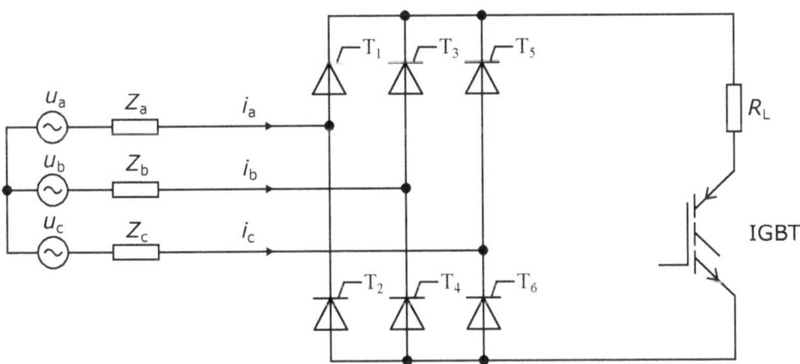

Figure 2. Impedance identification circuit for 20 kV medium voltage level.

For each measurement the following parameters are recorded over at least one period of 50 Hz:

- $i_a(t)/i_b(t)/i_c(t)$: current phase a/b/c
- $u_a(t)/u_b(t)/u_c(t)$: voltage phase a/b/c to earth

The loop impedances $\underline{Z}_{ab}(\omega)$, $\underline{Z}_{bc}(\omega)$, $\underline{Z}_{ca}(\omega)$ are derived from the recorded parameters with (1). These loop impedances can be rearranged into the line impedances $\underline{Z}_a(\omega)$, $\underline{Z}_b(\omega)$, and $\underline{Z}_c(\omega)$ [10]:

$$\underline{Z}_a(\omega) = \frac{1}{2} \cdot [\underline{Z}_{ab}(\omega) - \underline{Z}_{bc}(\omega) + \underline{Z}_{ca}(\omega)] \tag{2}$$

$$\underline{Z}_b(\omega) = \frac{1}{2} \cdot [\underline{Z}_{ab}(\omega) + \underline{Z}_{bc}(\omega) - \underline{Z}_{ca}(\omega)] \tag{3}$$

$$\underline{Z}_c(\omega) = \frac{1}{2} \cdot [-\underline{Z}_{ab}(\omega) + \underline{Z}_{bc}(\omega) + \underline{Z}_{ca}(\omega)] \tag{4}$$

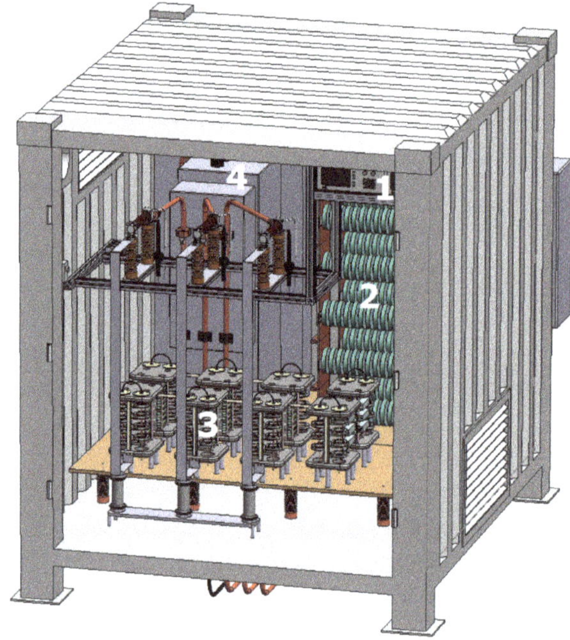

Figure 3. Impedance identification measurement device for 20 kV medium voltage level.

Typically, the impedances are average values over ten 50 Hz periods measured every five minutes over a month or more to identify impedance changes over daytime. Figure 4 shows a sample measurement. The compression approach, explained in Section 4, is applied to the voltage and current data, which is recorded during grid excitation and which is used for the calculation of the grid impedance. The authors want to determine the effect of data compression on the grid impedance calculation with this compressed raw dataset.

This setting yields 288 measurements per day or 8928 per month for each recorded voltage and current parameter. The sample rate is 500 kHz. Overall, the authors use a dataset with d = 206,744 data points (sample size d) for each recorded parameter, voltage and current. For an easy representation of the following figures and the compression dataset, the number of measuring points d was used (instead of the time vector t). Theoretically, the respective t-vector would have to be multiplied by the reciprocal of the sampling rate (500 kHz).

As the measurement device is remotely controlled and the data is transferred via the cellular network a compression is beneficial to optimize the data transfer speed and costs. Especially when the measurement device's installation site is in areas with low network coverage exhibiting low transfer rates. The transients in the recorded voltage and current parameters are the essential part of the data, as they determine the frequency dependency of the impedance values.

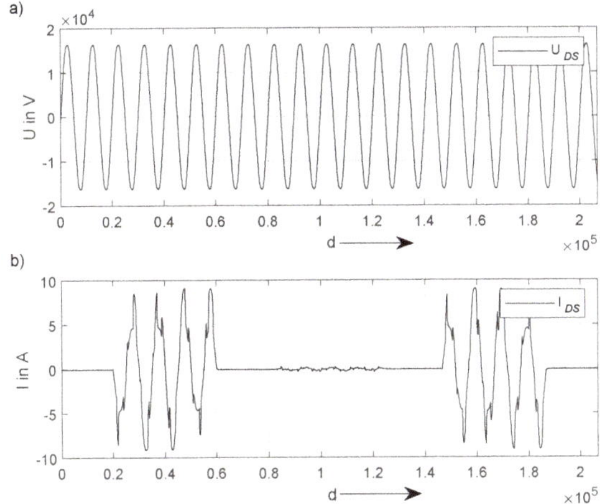

Figure 4. Measurement results with U_{DS} (**a**) and I_{DS} (**b**).

3. Lossy Compression Techniques

Various meta-analyses, types and overviews of data compression approaches can be found in [12–16]. The compression techniques are divided into lossy and lossless methods. Lossy ones generate better results by losing (preferably irrelevant) information. This can be explained by the fact that the result of the decompression is not identical to the starting dataset. In contrast, lossless methods produce an identical decompressed dataset [13].

The combination of impedance measurements and data compression can hitherto be found only in other fields of research. As an example serves the medical area where an extensive comparison of compression methods adapted to the impedance of cardiomyocytes is presented. The approach uses the wavelet transformation technique to analyse the effect of compression on sensitive data coming from cardiomyocytes and generating compression ratio of round about 5:1 [9].

Hereinafter follows a short description of the lossy compression methods that are compared in this paper. All of these approaches are frequently used for other types of data (SVD, WT) and appear interesting for the paper approach (TFA) [17–22]. The two well-known, widely used approaches WT and SVD are only briefly described. For further explanation, please consult the references that are listed in the Sections 3.1 and 3.2. TFA is explained in more detail but can be found in [17,22] if necessary.

3.1. SVD—Singular Value Decomposition

The so-called Singular Value Decomposition (SVD) splits a m × n set of data (voltage/current × time stamp) **DS** into three different matrices (5). The diagonal matrix $\mathbf{\Sigma}$ contains the singular values (SVs), see also Figure 5.

$$\mathbf{DS}_{m\times n} = \mathbf{U}_{m\times m}\mathbf{\Sigma}_{m\times n}\mathbf{V}^{\mathrm{T}}_{n\times n} \tag{5}$$

The data compression takes advantage of the fact that a close approximation of **DS** can be achieved by keeping the significant SVs of matrix $\mathbf{\Sigma}$. The compression success depends on the amount of reduction of singular values in $\mathbf{\Sigma}$.

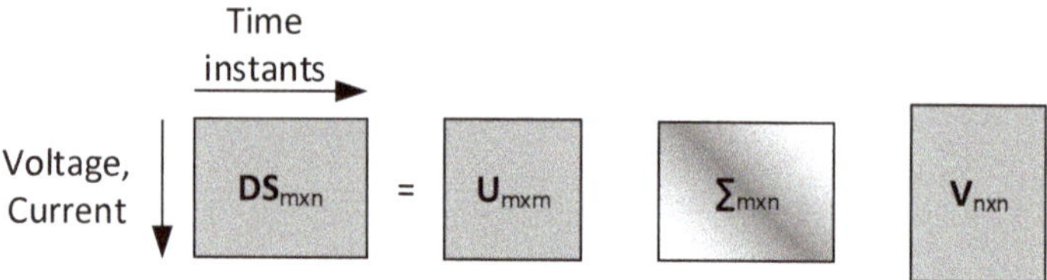

Figure 5. SVD of data matrix **DS**.

3.2. WT—Wavelet Transformation

A wavelet transform (WT) orthogonally decomposes a time series into wavelet and scaling coefficients. The main difference to the Fourier transform, which splits a signal into cosine and sine, is the use of (real and Fourier space) functions by the WT. The deletion of irrelevant data points increase the compression ratio and reduce the mean percentage error (MPE) and mean absolute error (MAE). That is why, it is important to find the best thresholds, levels of decomposition (LoD), and Daubechies' wavelets (DW). For further explanation, see [19–21].

3.3. TFA—Triangular Function Algorithm

The Triangular Function Algorithm (TFA) encloses the steps (I)-(VI) and is an enhanced version of the approach developed in [17]. (I) Read the dataset and choose your preferred percentiles (e.g., Q_5–Q_{95}). (II) Generate percentiles of the original dataset and save the data points $y_{i\,Q}$, $x_{i\,Q}$. Perform a moving average FIR filter to smooth the (remainder of the) dataset. (III) Read a_0 data points, which is the step width. (IV) Choose number of polynomials of least square fit. Perform Λ in (6), to obtain the slope b_1 and intercept b_0 (6). Determine the mean square error and unbiased standard deviation (σ).

$$\Lambda = \sum_{i=1}^{a} [y_i - (b_1 x_i + b_0)]^2 \quad (6)$$

In our case, quadratic or higher polynomial functions should be avoided because of the lower compression-error-ratio depending on the higher number of compressed and saved datapoints (e.g., b_2, b_1, b_0). (V) Read and check the following data point (y_i, x_i). If its value is within ($\pm m\sigma$, with factor m) the predicted values, jump to (III). Otherwise start a new line segment and go to step (IV). (VI) After compressing the whole dataset, insert percentiles (y_{iQ}, x_{iQ}) to finish the algorithm. For further explanation, please see [17].

4. Proposed Approach and Key Metrics

4.1. Novel Approach

The first step to compress and decompress the impedance data is the generation of the dataset obtained from impedance measurements (Section 2). In a second step, the dataset is smoothed to generate a periodic pulse signal with spikes. For this step, either the moving average filter (FIR filter) method or the Δ *sin*-signal method is chosen. The latter method uses the basic function of the voltage output (depending on the measured dataset (7)) to extract the noise of the input function of U_{DS}. The basic function (7) is determined by the impedance evaluation program and approximated by using fitting algorithm toolboxes.

$$U_{base} = \sqrt{2} \cdot 11.56\ \text{kV} \cdot \sin(2\pi \cdot 48\ \text{Hz} \cdot t) \quad (7)$$

The result of delta $(U_{base} - U_{DS})$ or the FIR-filter output from delta $(U_{FIR} - U_{DS})$ extracts the difference, the so-called spikes U_{noise}, I_{noise}, shown in Figure 6. These spikes are challenging to compress and the main reason for the complexity of the developed approach. Typically, existing programs (e.g., smooth from Matlab) smooth out these minimal data swings or spikes directly, thus eliminating the possibility to determine the grid impedance (depending on U_{DS}).

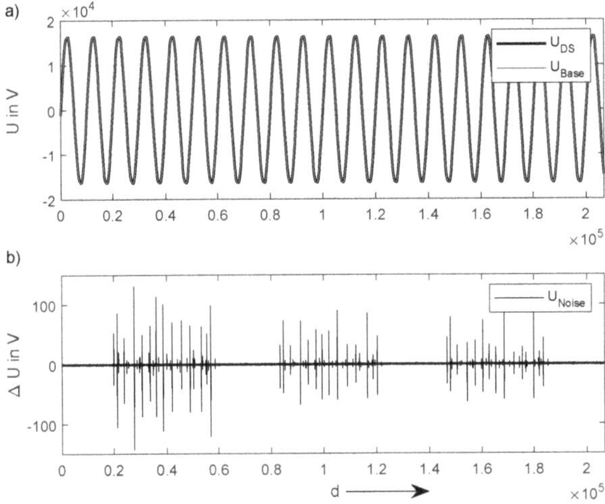

Figure 6. Overview of U_{DS} and generated U_{Base} (**a**) and U_{Noise} (**b**).

Regarding the current, this procedure can only be carried out with the FIR filter because the I function cannot be readjusted similarly.

Step 3 includes the compression of the dataset. Using different types of lossy compression algorithms, the spikes $U_{noise(C)}$ and the current $I_{DS(C)}$ are converted into a compressed dataset that is stored or saved in online or local data storage systems by their owner (e.g., distribution grid owner, utilities, etc.). After decompression, the recombination of the basic signal and the decompressed spikes is realised (only for the voltage). A validation of the output signal and an automatic performance check, including a comparison between the different compression approaches, ensues. As a result, the generated grid impedance can be compared to the input dataset (and the determined grid impedance) to determine the differences and whether or not the procedure has to be repeated. A flowchart and overview of these steps are shown in Figure 7.

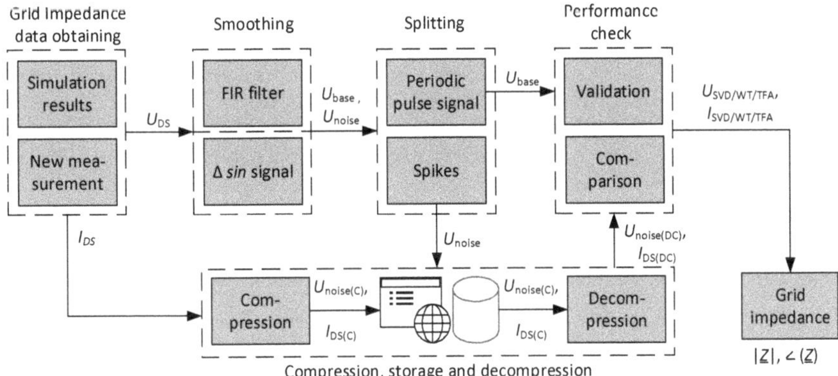

Figure 7. Flowchart of the novel paper approach.

4.2. Key Metrics

Between different compression algorithms, the compared key factors are the processing duration and the errors in combination with compression ratios to achieve an accurate reconstruction result. To ensure a valid comparison of the different types of compression algorithms, different key metrics are used. The compression ratio CR is defined in (8).

$$CR = \frac{\text{size of the input dataset (measured, uncompressed)}}{\text{size of the output dataset (compressed)}} \tag{8}$$

Following this definition, values greater than 1 indicate compression and values less than 1 imply expansion. The loss of information will be measured by comparing the reconstructed data matrix **XR** (with their rows and columns $n_{\text{row}} \cdot n_{\text{col}}$) with the original data matrix **X**. The so-called MAE—mean absolute error is defined in (9).

$$MAE = \frac{1}{n_{\text{row}} \cdot n_{\text{col}}} \sum_{i=1}^{n_{\text{row}}} \sum_{j=1}^{n_{\text{col}}} |\mathbf{X}(i,j) - \mathbf{XR}(i,j)| \tag{9}$$

5. Results

To generate comparable results, all compression approaches are set on a CR round about 4:1. This CR is like a trade-off between the advantages of compression and sufficiently high data fidelity, but randomly chosen for this test case. Although higher CRs are technically possible, this is not the objective of this paper. A comparison of all results for the compression of U_{Noise} and I_{DS} is displayed in Table 1.

Table 1. Compression factor, MAE and processing time of different compression approaches for U_{noise} and I_{DS}.

Type	CR (U)	MAE (U)	t (U)	CR (I)	MAE (I)	t (I)
SVD	4.1:1	0.36	120 s	4.2:1	0.001	113 s
WT	4.0:1	0.39	8 s	4.0:1	0.002	31 s
TFA	4.2:1	0.29	101 s	4.7:1	0.17	261 s

The proposed approaches have been implemented using MATLAB and performed on a PC Intel core i5-3210M processor, 2.50 GHz, with 4 GB of RAM. To illustrate the difference between the approaches, Figures 8 and 9 show the resulting decompression graphs for U_{Noise} and I_{DS}. In particular, the TFA algorithm does not possess good compression properties, especially when looking at Figure 9. The original curve is simply linearized because of a sluggish (high) threshold ($\pm m\sigma$), even if the resulting MAE is as good as in the other compression methods.

Only the WT and SVD algorithms yield accurate values for the impedance after the recombination of U_{base} and the decompressed $U_{\text{noise(DC)}}$ including the decompressed current $I_{\text{DS(DC)}}$. WT shows a particularly good fit of its decompressed impedance values Z to the original values Z_1 (Figure 10). Based on (10), Figure 10 shows the absolute impedance $|Z|$ (a), the impedance angle $\angle(Z) = \varphi_Z$ (b), and the absolute deviation $\Delta\,|Z1 - Z|$ (c) over the frequency f.

$$\underline{Z} = |Z|\,\varphi_Z \tag{10}$$

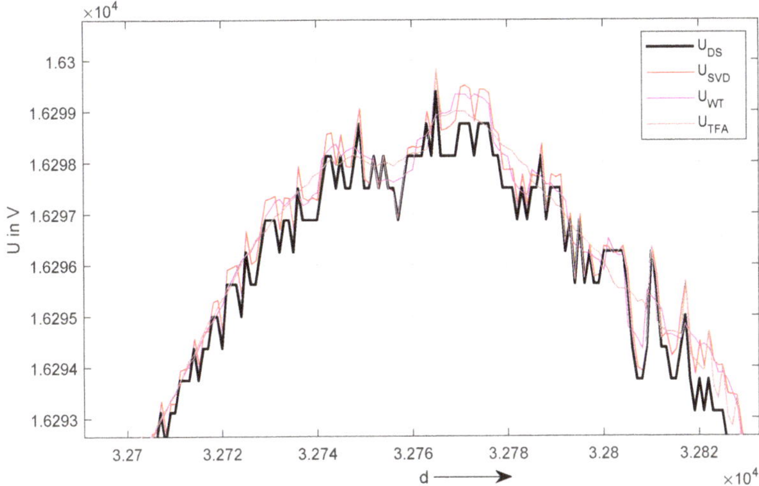

Figure 8. Comparison of U_{DS} and the resulting compression outputs $\sum(U_{base}, U_{noise(DC)})$ of the different lossy approaches.

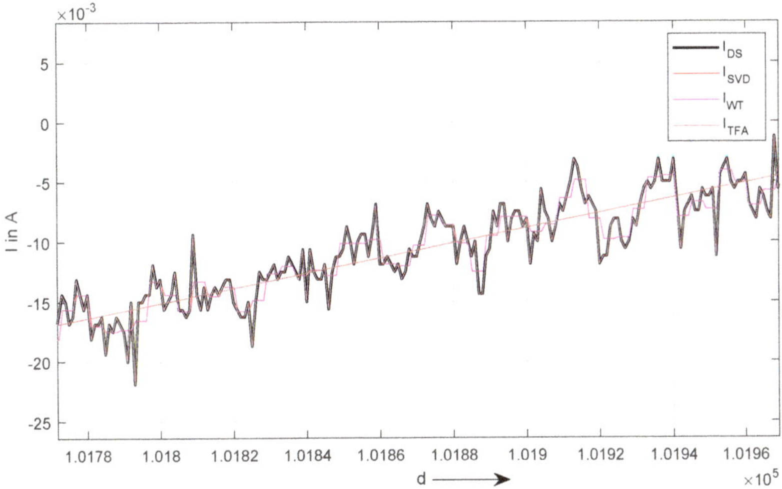

Figure 9. Comparison of I_{DS} and the resulting compression output of the different lossy approaches $I_{DS(DC)}$.

Only for frequencies $\geq$ 25 kHz do the discrepancies in the WT results increase slightly ($\geq$1%), see Figure 10 (bottom). In comparison see Figure 11, the results obtained by the SVD algorithm deviate from the original dataset by almost 2% for f $\geq$ 25 kHz. Depending on I_{DS}, the TFA algorithm produces a large deviation of the MAE that leads to the significantly worse results.

Additionally, both the SVD and the TFA algorithms show high processing times $t(I), t(U)$ (Table 1). An evaluation of the best fitting technique based on the processing time is only possible to a limited extent. Compression using SVD and WT should be investigated for each data set separately. Conclusively, the TFA algorithm is deemed inadequate to handle the task discussed in this paper or similar tasks.

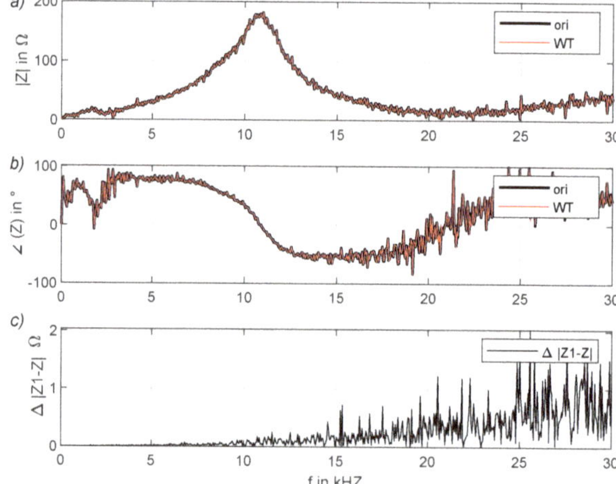

Figure 10. Impedance measurement results of the original (ori = original dataset and their impedance Z_1) and decompression results using WT (WT = WT decompressed dataset and their impedance Z). The absolute impedance $|Z|$ (**a**), the impedance angle $\angle(Z) = \varphi_Z$ (**b**), and the absolute deviation $\Delta\,|Z1 - Z|$ (**c**) over the frequency f are shown.

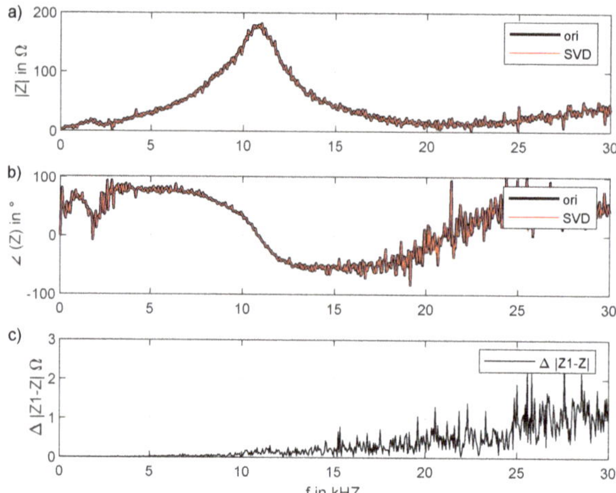

Figure 11. Impedance measurement results of the SVD, for explanation see Figure 10.

6. Conclusions

A more detailed understanding of the effects of different voltage and current profiles on the grid impedance requires large amounts of data, especially in the future.

Thus, a very important question is whether or not it is possible to compress data sets from impedance measurements of energy systems by using lossy compression algorithms while maintaining data fidelity. In this paper, the authors show that especially the WT (Wavelet transform) show promising results by reducing the size of the dataset in an efficient way without losing relevant information. The SVD (Singular Value Decomposition) generates almost comparable results, but need much more processing time, referring to the used programming language and computational environment. The presented approach allows of easy and effective data compression with only limited computational resources and, as a result, an increase in the number of measurements that can be stored. It is also conceivable that the presented approach can be applied for efficient online or on-site external server data backup.

While high compression levels are useful in order to reduce the data amount for e.g., utilities, the needed level of data fidelity in the output dataset depends on the targeted application. That is why it is not the objective of this paper to create the best technique and perform on a given dataset. Lossy compression works better when the nature of the compressed data is taken into account (e.g., such as human ear characteristics in MP3). To be able to determine the limits of usability of lossy compression methods, further analyses need to be done. It must be analysed which lossy method generates the best possible outcome i.e., the maximum level of accuracy with the highest suitable compression ratio.

A comparison with lossless compression methods is also interesting. And the authors want to assess if the proposed method is suitable for on-line impedance measurement [4,5].

Author Contributions: M.P. modeled and validated the approaches via simulation topology, data analysis; prepared original. M.F.M. modeled the mid voltage impedance measurement system. D.S. and M.M. contributed to review and editing. F.G. developed and validated one compression method. All of them also supported the draft preparation. D.B. proofread and validate the re-submission. M.M. and D.S. provided resources and supervision. All authors have read and agreed to the published version of the manuscript.

Funding: This research was founded by the Federal Ministry for Economic Affairs and Energy in the course of the project "NEW 4.0—Norddeutsche EnergieWende" (Northern German Energy Transition). The support code is 03SIN414.

Conflicts of Interest: The authors declare no conflict of interest.

Abbreviations

The following abbreviations are used in this manuscript:

CR	Compression Ratio
FIR	Finite Impulse Response
IGBT	Insulated Gate Bipolar Transistor
MAE	Mean Absolute Error
NIM	Network Impedance Measurement
PCC	Point of Common Coupling
PEC	Power Electronic Components
PWM	Pulse Width Modulation
RFA	Rectangular Function Algorithm
SVD	Singular Value Decomposition
TFA	Triangular Function Algorithm
WT	Wavelet Transform

References

1. Blaabjerg, F.; Yang, Y.; Yang, D.; Wang, X. Distributed Power-Generation Systems and Protection. *Proc. IEEE* **2017**, *105*, 1311–1331. [CrossRef]
2. Azzouz, M.A.; El-Saadany, E.F. Multivariable Grid Admittance Identification for Impedance Stabilization of Active Distribution Networks. *IEEE Trans. Smart Grid* **2017**, *8*, 1116–1128. [CrossRef]
3. Jordan, M.; Grumm, F.; Kaatz, G.; Meyer, M.F.; Wilken, H.; Schulz, D. Online Network Impedance Spectrometer for the Medium-Voltage Level. In Proceedings of the 2018 IEEE International Conference on Environment and Electrical Engineering and 2018 IEEE Industrial and Commercial Power Systems Europe (EEEIC/ICPS Europe), Palermo, Italy, 12–15 June 2018.
4. Eskandari, M.; Li, L.; Moradi, M.; Siano, P.; Blaabjerg, F. Optimal Voltage Regulator for Inverter Interfaced Distributed Generation Units Part: Control System. *IEEE Trans. Sustain. Energy* **2020**. [CrossRef]
5. Eskandari, M.; Li, L.; Moradi, M.; Siano, P.; Blaabjerg, F. Optimal Voltage Regulator for Inverter Interfaced Distributed Generation Units Part: Application. *IEEE Trans. Sustain. Energy* **2020**. [CrossRef]
6. Xin, H.; Li, Z.; Dong, W.; Wang, Z.; Zhang, L. A Generalized-Impedance Based Stability Criterion for Three-Phase Grid-Connected Voltage Source Converters. *arXiv* **2017**, arXiv:1703.10514.

7. Wen, B.; Dong, D.; Boroyevich, D.; Burgos, R.; Mattavelli, P.; Shen, Z. Impedance-based analysis of grid-synchronization stability for three-phase paralleled converters. *IEEE Trans. Power Electron.* **2016**, *31*, 26–38. [CrossRef]
8. Sumner, M.; Palethorpe, B.; Thomas, D. Impedance Measurement for Improved Power Quality—Part 1: The Measurement Technique. *IEEE Trans. Power Deliv.* **2004**, *19*, 1442–1448. [CrossRef]
9. Guyot, P.; Batista, L.; Djermoune, E.H.; Moureaux, J.M.; Bastogne, T.; Doerr, L.; Beckler, M. Comparison of compression methods for impedance and field potential signals of cardiomyocytes. In Proceedings of the 2017 Computing in Cardiology (CinC), Rennes, France, 5 April 2018.
10. Do, T.T. *Measurement Device for Mobile Identification of the Grid Impedance*; VDE-Verlag: Berlin, Germany, 2014; ISBN 978-38007-3633-1.
11. Wilken, H.; Jordan, M.; Schulz, D. Spectral Grid Impedance Identification on the Low-, Medium-and High-Voltage Level–System Design, Realization and Measurement Results of Grid Impedance Measurement Devices. *Adv. Sci. Technol. Eng. Syst. J.* **2019**, *4*, 08–16. [CrossRef]
12. Storer, J. *Data Compression: Methods and Theory*; Computer Science Press: New York, NY, USA, 1988.
13. Salomon, D.; Motta, G. *Handbook of Data Compression*, 5th ed.; Springer Science & Business Media: London, UK, 2010.
14. Blelloch, E. *Introduction to Data Compression*; Computer Science Department, Carnegie Mellon University: Pittsburgh, PA, USA, 2010.
15. Pu, I. *Fundamental Data Compression*; Butterworth-Heinemann: London, UK, 2005.
16. Tate, J. Preprocessing and Golomb–Rice Encoding for Lossless Compression of Phasor Angle Data. *IEEE Trans. Smart Grid* **2016**, *7*, 718–729. [CrossRef]
17. Clements, A.; McCulloch, M.; Nixon, K. Low-loss, high-compression of energy profiles. In Proceedings of the Renewable Energy Research and Applications (ICRERA), Palermo, Italy, 22–25 November 2015.
18. de Souza, C.; Assis, T.; Pal, B. Data compression in smart distribution systems via singular value decomposition. *IEEE Trans. Smart Grid* **2017**, *8*, 275–284. [CrossRef]
19. Wen, L.; Zhou, K.; Yang, S.; Li, L. Compression of smart meter big data: A survey. *Renew. Sustain. Energy Rev.* **2018**, *91*, 59–69. [CrossRef]
20. Engel, D. Wavelet-based Load Profile Representation for Smart Meter Privacy. In Proceedings of the IEEE PES Innovative Smart Grid Technologies (ISGT'13), Washington, DC, USA, 24–27 February 2013.
21. Engel, D.; Eibl, G. Wavelet-Based Multiresolution Smart Meter Privacy. *IEEE Trans. Smart Grid* **2016**, *99*, 1–12. [CrossRef]
22. Plenz, M.; Dong, C.; Grumm, F.; Meyer, M.F.; Schumann, M.; McCulloch, M.; Jia, H.; Schulz, D. Framework Integrating Lossy Compression and Perturbation for the Case of Smart Meter Privacy. *Electronics* **2020**, *9*, 465. [CrossRef]

Article

Security Assessment and Coordinated Emergency Control Strategy for Power Systems with Multi-Infeed HVDCs

Qiufang Zhang, Zheng Shi, Ying Wang *, Jinghan He, Yin Xu and Meng Li

School of Electrical Engineering, Beijing Jiaotong University, Beijing 100044, China; 16117384@bjtu.edu.cn (Q.Z.); shizheng@bjtu.edu.cn (Z.S.); jhhe@bjtu.edu.cn (J.H.); xuyin@bjtu.edu.cn (Y.X.); mengl@bjtu.edu.cn (M.L.)

* Correspondence: yingwang1992@bjtu.edu.cn; Tel.: +86-1520-131-5768

Received: 23 May 2020; Accepted: 15 June 2020; Published: 19 June 2020

Abstract: Short-circuit faults in a receiving-end power system can lead to blocking events of the feed-in high-voltage direct-current (HVDC) systems, which may further result in system instability. However, security assessment methods based on the transient stability (TS) simulation can hardly catch the fault propagation phenomena between AC and DC subsystems. Moreover, effective emergency control strategies are needed to prevent such undesired cascading events. This paper focuses on power systems with multi-infeed HVDCs. An on-line security assessment method based on the electromagnetic transient (EMT)-TS hybrid simulation is proposed. DC and AC subsystems are modeled in EMTDC/PSCAD and PSS/E, respectively. In this way, interactions between AC and DC subsystems can be well reflected. Meanwhile, high computational efficiency is maintained for the on-line application. In addition, an emergency control strategy is developed, which coordinates multiple control resources, including HVDCs, pumped storages, and interruptible loads, to maintain the security and stability of the receiving-end system. The effectiveness of the proposed methods is verified by numerical simulations on two actual power systems in China. The simulation results indicate that the EMT-TS hybrid simulation can accurately reflect the fault propagation phenomena between AC and DC subsystems, and the coordinated emergency control strategy can work effectively to maintain the security and stability of systems.

Keywords: receiving-end system; multi-infeed HVDCs; security assessment; emergency control strategy; electromagnetic transient (EMT)-transient stability (TS) hybrid simulation

1. Introduction

With the growing penetration of line-commutated converter-based high-voltage direct-current (LCC-HVDC) lines, power systems with multi-infeed HVDCs, where several HVDC lines feed into nearby AC systems, are becoming more common [1,2]. Due to the complicated interactions among HVDCs and AC systems, such systems are facing challenges in secure and stable operation, especially when the short-circuit capacity of the receiving-end AC system is low relative to the rated power of the HVDCs [3–5]. An AC system fault that occurs at the receiving-end system can cause not only commutation failure of the directly-connected HVDC but also concurrent commutation failures or even blockings of adjacent HVDCs, giving rise to risks of instability and large-scale blackouts [6–8]. Therefore, it is critical to conduct on-line pre-decisions before such credible contingencies occur so that effective emergency controls can be implemented in time to prevent such cascading failures.

There are two steps involved in the on-line pre-decision-making [9–11]. One is the security assessment, which estimates the system security and stability under anticipated contingencies at the current operation point. The other is emergency control strategy decision-making, which generates emergency control strategies based on the security assessment result. Therefore, a control strategy table composed of emergency control strategies and corresponding contingencies will be generated in the pre-decision-making. Once a contingency occurs, emergency controls can be implemented in time by searching the control strategy table. In the on-line pre-decision-making, the control strategy table is updated within a fixed period to adapt to the changing operating conditions.

Up to now, many security assessment methods have been proposed for AC/DC systems. The time-domain simulation method is widely used for its good model extensibility and can be classified into two categories: One is based on mature transient stability (TS) simulators with the built-in models and solvers, like Power System Simulator/Engineering (PSS/E) [12], Bonneville Power Administration (BPA) [13], and Transient Security Assessment Tool (TSAT) [14], and the other is based on customized models and solving algorithms, like the voltage source equivalent-based method [15], multi-decomposition method [16], and optimal subinterval selection method [17]. However, in these methods, HVDC converters are expressed by steady-state models, and the fault propagation phenomena between AC and DC subsystems, such as commutation failures and blocking events caused by AC system faults, may not be reflected accurately. Similarly, in transient energy-based methods [18,19] and their derived methods, which combine them with time-domain simulation methods [20,21], the transient energy function cannot incorporate HVDC converter-involved dynamics, and there is a probability that the commutation failures or blocking event-related issues cannot be identified. However, considering the credible impact of interactions between AC and DC subsystems on the secure and stable operation, accurately detecting the fault propagation phenomena is crucial in the above methods [22,23]. Recently, data-driven artificial intelligence (AI) methods have been proposed as fast tools, e.g., the generative adversarial network (GAN) [24], convolutional neural network (CNN) [25], and deep belief network (DBN) [26]. Most of these methods are at their early stages and their practicality needs to be improved [26]. Therefore, improving the accuracy of the time-domain simulation method or transient energy-based methods is necessary. In fact, to describe the detailed dynamics of HVDCs accurately, electromagnetic transient (EMT) simulation is a suitable tool, but it cannot be used directly in the on-line security assessment due to its low computational efficiency [27]. Therefore, a method that can take advantage of the modeling accuracy of EMT and the computational efficiency of existing security assessment methods should be explored. EMT-TS hybrid simulation, in which the HVDC-related subsystems are modeled in EMT and the rest in TS, provides an idea for solving the problem.

In emergency control, load shedding (LS) is a common measure and its optimization method is continuously improved to achieve cost-effective control for issues like frequency instability [28] and voltage collapse [29]. Subsequently, considering that large disturbances can affect the power angle, voltage, and frequency simultaneously, the authors of [30] constructed an LS optimization model considering multiple security constraints, including transient voltage deviation security, transient frequency deviation security, and transient angle stability, which can remedy the limitation of single security constraint-based methods. In addition to LS, other control resources, such as HVDCs [31,32] and pumped storages [33], can also be used for emergency control. However, their control amount is usually determined separately [31–33]. The authors of [34,35] comprehensively coordinate HVDCs, pumped storages, and interruptible loads in the emergency control strategy to handle frequency stability issues in the East China power grid. Nevertheless, similar to [28], only frequency instability is considered in the proposed scheme. The authors of [36] developed a multi-resource coordinated control strategy for an actual power grid to cope with the impact the DC blockings have on weak AC channels, but it was obtained based on the characteristics of the grid without mathematical analysis, which may be not suitable for other grids.

According to the above analysis, for power systems with multi-infeed HVDCs: (1) A security assessment method that can well reflect the fault propagation phenomena between AC and DC subsystems, and generate reliable results within an acceptable time should be studied; and (2) the emergency control strategy that can comprehensively coordinate multiple control resources while satisfying multiple critical security constraints is needed.

In this paper, an on-line pre-decision-making scheme, including security assessment and emergency control strategy decision-making, is proposed for power systems with multi-infeed HVDCs. The contributions are as follows:

(1) A security assessment method based on EMT-TS hybrid simulation is achieved. DC and AC subsystems are modeled in EMTDC/PSCAD and PSS/E, respectively. The security assessment method can accurately identify the security and stability issues related to interactions between AC and DC subsystems while maintaining the high computational efficiency;
(2) An emergency control strategy decision-making method that can coordinate HVDCs, pumped storages, and interruptible loads is developed subject to multiple security constraints. The decision-making method can minimize the control costs while maintaining the security and stability of the receiving-end system.

This paper is structured as follows: Section 2 introduces the procedure of the on-line pre-decision-making scheme. Section 3 describes the implementation of the security assessment based on EMT-TS hybrid simulation. Section 4 presents the optimization model and solution method of the emergency control decision-making problem. Two actual provincial systems in China are used to verify the proposed method in Section 5. Section 6 concludes the paper.

2. Procedure of the On-Line Pre-Decision-Making Scheme

In the on-line decision-making scheme, the control strategy table is updated at fixed intervals. During each interval, the operating condition of the system is assumed as being unchanged [11], and the anticipated contingencies include merely the fault and protection action information. According to the severity and probability, the contingencies can be divided into three levels [37]: (1) Single component fault; (2) single severe fault; and (3) multiple severe faults. Especially, in the third level, operation failure of the protection and reclosing failure caused by a permanent fault may induce HVDC blocking events and result in instability of the receiving-end system [38], which should be paid more attention to.

When updating the control strategy table, security assessment is conducted for the anticipated contingency set based on the current operating condition, and the emergency control strategy will be developed if system security and stability issues arise. Therefore, the procedure can be divided into three stages, as shown in Figure 1.

(1) Off-line preparation. Construct the EMT-TS hybrid model based on the information of the network topology, electrical parameters, control parameters, etc. Then, generate an off-line control strategy table under the anticipated contingency set and pre-determined typical operating conditions (different from the on-line control strategy table, various typical operating conditions need to be considered in the off-line control strategy table [39]), which will provide the initial solution of the decision-making model for the emergency control strategy;
(2) On-line security assessment based on EMT-TS hybrid simulation. Update the real-time operating state data, including the operation mode of the system and the power flow of the main transmission section; choose one contingency from the anticipated contingency set and run the hybrid simulation. Then, identify possible security and stability issues according to the security indices. Finally, generate the assessment result for contingencies that cause security and stability issues, including the current operating condition, contingency, and power shortage in the receiving-end system; and
(3) Emergency control strategy decision-making. Initialize the decision-making model with the operating condition and the control strategy. The operating condition is obtained from the assessment result. The control strategy, which is used as the initial solution, is determined based on the power shortage and the control strategy obtained through an approximate search of the off-line control strategy table. Then, solve the decision-making model based on the beetle antennae search (BAS) algorithm [40], a meta-heuristic algorithm developed by the inspiration of the beetle forging principle, until the termination criteria are met.

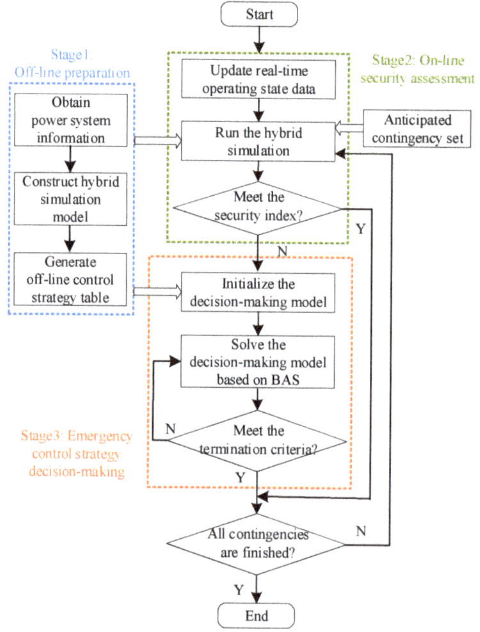

Figure 1. Procedure of the on-line pre-decision-making scheme.

3. Security Assessment Based on EMT-TS Hybrid Simulation

Security assessment refers to the analysis required to determine whether a power system can meet specified security criteria in both transient and steady-state time frames under credible contingencies [41]. Therefore, assessment methods and security indices are two of the parts involved in the security assessment. Considering that commutation failures and blocking events caused by AC system faults are typical fault propagation phenomena between AC and DC subsystems, the analysis of commutation failures and blocking events simulation is firstly analyzed in the following subsections. Then, the principle of EMT-TS hybrid simulation modeling and the security assessment index system are introduced.

3.1. Analysis of Commutation Failures and Blocking Events Simulation

The essence of the commutation failure is that the thyristor cannot establish a forward voltage blocking capability due to the insufficient negative voltage time, which can be represented by the extinction angle [42]. Therefore, a commutation failure can be considered to occur when the extinction angle is less than the inherent limit of the thyristor. As stated in [38], a commutation failure, which occurs again after an interval of 200 ms, is called a continuous commutation failure in engineering and may cause an HVDC blocking event. Therefore, in the study, a continuous commutation failure with an interval of 200 ms is taken as the condition of HVDC blocking.

However, in the simulation analysis, different criteria are developed to determine the occurrence of commutation failures and blocking events due to different modeling methods of HVDC converters. Table 1 compares the typical criteria of commutation failures and blocking events in the pure TS simulation and EMT-TS hybrid simulation. In the pure TS simulation, the models of the HVDC converter, such as the CDC4 model in PSS/E, are represented by steady-state equations. That is, the HVDC converter is modeled without thyristor valves, so commutation failures and blocking events can only be identified according to the AC voltages at commutation buses [43]. The AC voltage criteria are usually obtained under the assumption of an infinite AC system and the effect of voltage waveform distortion on commutation failures is ignored, so the accuracy is poor [42]. In the EMT-TS hybrid simulation, HVDC converters are modeled by thyristor valves, which are consistent with the actual condition, so commutation failures and blocking events can be identified accurately through detection of the extinction angle and the interval between two commutation failures.

Table 1. Typical criteria of commutation failures and blocking events in two simulation methods.

Simulation Methods	Commutation Failures	Blocking Events
Pure TS simulation	AC voltage at the inverter side (e.g., 0.785 p.u.)	AC voltage at the rectifier side (e.g., 0.6 p.u.)
EMT-TS simulation	Extinction angle (7.2° [44])	Interval between two commutation failures (200 ms [38])

Therefore, the EMT-TS hybrid simulation can achieve more accurate results in the commutation failures and blocking events simulation. It is more suitable for the security assessment of receiving-end systems to identify HVDC-related security and stability issues, which is validated in Section 5.

3.2. Principle of EMT-TS Hybrid Simulation Modeling

To build the hybrid simulation platform, two mature business software, PSS/E [45] and EMTDC/PSCAD [46], are integrated based on the interface software E-Tran Plus [47]. To construct the hybrid simulation model, several issues should be addressed:

(1) Interface location. As shown in Figure 2, the power system will be divided into two parts: The internal network and the external network. The internal network is comprised of HVDCs and the nearby AC buses, and it is modeled in the EMT simulator EMTDC/PSCAD. The rest of the system is the external network and is represented in the TS simulator PSS/E. To guarantee the accuracy and efficiency of the hybrid simulation, a proper interface location should be identified.

(2) Equivalent models of the external and internal networks. For the model in EMTDC/PSCAD, in addition to the detailed model of the internal network, an equivalent model of the external network needs to be constructed to ensure the integrity of the system. Similarly, an equivalent model of the internal network in PSS/E is also indispensable.

(3) Interaction protocol and data. During the hybrid simulation, the two simulators will exchange data through a certain interaction protocol to update the states of the equivalent models in time.

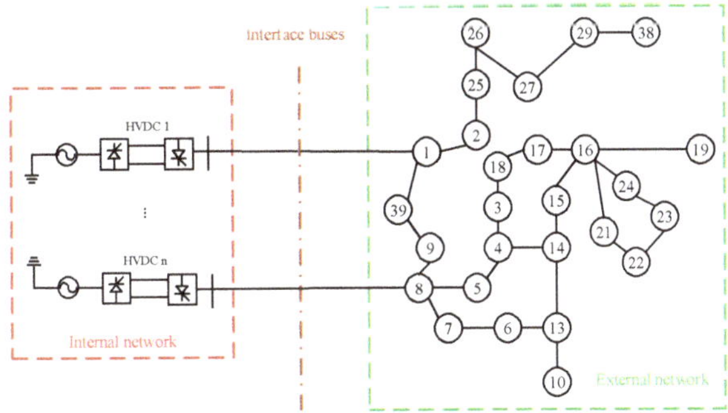

Figure 2. Topology of the power system.

3.2.1. Identify the Interface Location

When HVDC was first simulated in an EMT-TS hybrid simulation, the interface was located at the terminal buses of converters [48,49]. Subsequently, considering that TS simulation based on the fundamental frequency positive-sequence phasor model cannot effectively represent the waveform distortion or phase imbalance at converter terminals, an extension of the internal network into the AC system was suggested [50]. However, the specific methods for identifying the interface location were not mentioned. In PSS/E, phase imbalance caused by asymmetrical faults can be described by appending negative-sequence and zero-sequence parameters to the positive-sequence system, so the interface location mainly depends on the description of harmonic distortion, which is related to the frequency [51].

Based on the above analysis, a frequency-domain characteristics analysis method is used here to identify the location of the interface. The range of the internal network is expanded continuously and the impedance-frequency characteristics at the buses of interest are analyzed in the hybrid simulation, until the differences among the impedance-frequency characteristics under different locations reduce to a certain range. That is, expanding the scope of the internal network has almost no effect on the impedance-frequency characteristics anymore. Then, the interface location is finally identified based on the smaller internal network of the last two-scope internal networks.

In the security assessment of power systems with multi-infeed HVDCs, HVDC dynamics are essential and should be described accurately. Therefore, the commutation buses at the rectifier side and inverter side can be taken as the buses of interest.

3.2.2. Equivalent Models of the External and Internal Networks

In the study, the construction of equivalent models is implemented in E-Tran Plus. In order to consider the asymmetrical faults, a multi-port three-phase equivalent circuit with voltage sources, PI sections and transformers, is constructed in EMTDC/PSCAD to represent the external network. PI sections represent the impedance between buses of the same voltage level, whereas transformers represent the impedance between buses of different voltage levels. As for the equivalent model of the internal network, the generator model is used in PSS/E. When performing the power flow calculation to get the updated data, which will be transferred to EMTDC/PSCAD, the generator model will act as a current injection and a change in the system admittance matrix in PSS/E. The equivalent models of both networks can be found in the implementation of the hybrid simulation shown in Figure 3.

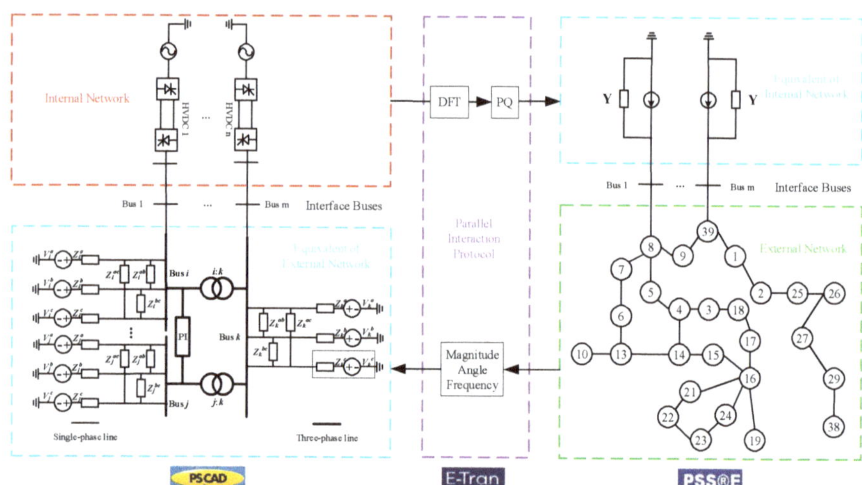

Figure 3. Implementation of the hybrid simulation.

3.2.3. Interaction Protocol and Data

A parallel interaction protocol is adopted to exchange the updated data, indicating both simulators run simultaneously during the simulation process. Before the simulation, initialization will be executed, in which the equivalent voltage sources in EMTDC/PSCAD and the equivalent generators (or current sources and admittance matrix) in PSS/E are initialized based on the power flow results of the pure TS simulation in PSS/E. During the simulation, the voltage magnitude, phase angle, and frequency information from PSS/E will be sent to EMTDC/PSCAD to update the equivalent voltage sources. At the same time, a discrete Fourier transform (DFT) will be used to extract PQ values from EMTDC/PSCAD to update the equivalent generators. All the data are exchanged at the time step of the TS simulation.

3.3. *Security Assessment Index System*

During the security assessment, the EMT-TS hybrid simulation model is updated with the real-time operating data obtained by the intelligent measurement system and run under the pre-defined contingency. Then, the results are evaluated based on a security assessment index system to identify the security and stability issues. Once any security index in the simulation results exceeds the preset range, the current operating condition, contingency, and power shortage in the receiving-end system will be sent to the decision-making model to obtain the optimal emergency control strategies.

The security assessment index system is composed of static security indices and dynamic security indices. The steady-state frequency deviation, voltage deviation, and power flow of the lines belong to static indices while the maximum/minimum value of the transient voltage and frequency, as well as the maximum transient relative power angle, belong to dynamic indices. Referring to [52], the preset ranges of the security assessment index system are shown in Table 2. In static indices, the threshold values of the steady-state frequency deviation Δf and steady-state voltage deviation ΔV are 0.05 Hz and 0.1 p.u., respectively; and the power flow of lines should be less than the transmission power limit $p_{\max}$, which is 1 p.u. in the study. As for dynamic indices, the security threshold of equipment, as well as coordination among different controls, needs to be considered. To ensure the safety of power system equipment, the maximum value of the transient voltage should be less than 1.3 p.u.; to avoid triggering low-voltage LS, high-frequency generator tripping, and low-frequency LS, the minimum value of the transient voltage should be higher than 0.85 p.u. and the threshold values of the maximum/minimum transient frequency are 51.5 and 49.25 Hz, respectively. At the same time, the power angle difference $\Delta\delta$ of any two units should be less than 360° to avoid the out-of-step of the first and second pendulums.

Table 2. Preset ranges of the security assessment index system.

Static Security Indices	Preset Range	Dynamic Security Indices	Preset Range
steady-state frequency deviation (Hz)	$\|\Delta f\| < 0.05$	maximum/minimum transient frequency (Hz)	$49.5 < f < 50.5$
steady-state voltage deviation (p.u.)	$\|\Delta V\| < 0.1$	maximum/minimum transient voltage (p.u.)	$0.85 < V < 1.1$
steady-state power flow of lines (p.u.)	$p < p_{\max}$	maximum transient relative power angle (°)	$\Delta\delta < 360°$

4. Emergency Control Strategy Decision-Making Based on BAS

When a security or stability issue is identified by security assessment, the emergency control strategy will be generated by solving the decision-making model with BAS. In the following subsections, the mathematical decision-making model and the decision-making procedure of the emergency control strategy are described.

4.1. *Mathematical Decision-Making Model*

The emergency control strategy decision-making problem can be formulated as a constrained optimization problem. The objective includes minimizing control costs and deviations of the frequency and voltage, and adjustment amount constraints, steady-state constraints, and transient-state constraints are considered.

4.1.1. Objective Function

The primary objective is minimizing the total control costs of multiple resources, and the secondary objective is minimizing the total weighted deviations of the frequency and voltage. The two objectives are normalized and combined through a weighted coefficient to formulate the objective function f, as shown in the following equations:

$$\min f = f_1 + \omega_0 f_2, \tag{1}$$

$$\min f_1 = \left(\sum_{i=1}^{N_{\mathrm{D}}} \Delta p_i^{\mathrm{DC}} + \sum_{j=1}^{N_{\mathrm{S}}} \Delta p_j^{\mathrm{pump}} \cdot x_j + \sum_{k=1}^{N_{\mathrm{L}}} \Delta p_k^{\mathrm{load}}\right) / S^{\mathrm{base}}, \tag{2}$$

$$\min f_2 = \varepsilon_f \sum_g \frac{\Delta f_g(p)}{\sigma_g f^{\mathrm{base}}} + \varepsilon_v \sum_m \frac{\Delta V_m(p)}{V^{\mathrm{base}}} K_V, \tag{3}$$

$$\Delta f_g(p) = \Delta f^{\mathrm{s}}(p) + \Delta f_g^{\mathrm{d}}(p), \tag{4}$$

$$\Delta V_m(p) = \Delta V_m^{\mathrm{s}}(p) + \Delta V_m^{\mathrm{d}}(p), \tag{5}$$

where f_1 is the primary objective; f_2 is the secondary objective; ω_0 is the weighted coefficient; N_{D}, N_{S}, and N_{L} are the numbers of HVDCs, pumped storages, and interruptible loads in emergency resources; Δp_i^{DC} is the power adjustment of HVDC i; $\Delta p_j^{\mathrm{pump}}$ is the consumed power of the tripped pumped storage j; x_j is a 0–1 variable, 1 represents tripping the pumped storage j while 0 represents keeping the original state; $\Delta p_k^{\mathrm{load}}$ is the power adjustment of load k; S^{base} is the base value of the power system capacity; ε_f and ε_v are the weighted coefficients of the frequency and voltage; Δf_g and σ_g are the frequency deviation and coefficient of the primary frequency adjustment at the generator g, respectively; f^{base} is the reference frequency; ΔV_m is the voltage deviation of bus m; K_V is the voltage regulation factor; V^{base} is the reference voltage; Δf^{s} is the steady-state frequency deviation of the system; Δf_g^{d} is the transient-state frequency deviation of bus g; and ΔV_m^{s} and ΔV_m^{d} are the steady-state and transient-state voltage deviations of bus m.

Equation (1) is the objective function, in which the weighted coefficient ω_0 is defined by users. Equation (2) is the primary objective, with Δp_i^{DC}, x_j, and $\Delta p_k^{\mathrm{load}}$ as decision variables. Equation (3) describes the secondary objective. Considering that the power imbalance of the receiving-end system due to the HVDC blocking event will seriously affect the system frequency, assume $\varepsilon_f > \varepsilon_v$. Equations (4) and (5) represents the frequency deviation and voltage deviation, respectively.

It should be noted that the priority of three kinds of control resources is different, which is reflected by the control action time in the control strategy. Taking the control speed and control cost into account, the action sequence adopted here is HVDCs, pumped storages, and interruptible loads. Considering the communication delay and control device response time, the control action time of HVDCs is 100 ms after the security or stability issue occurs, and the pumped storages and interruptible loads are followed, which are 300 and 500 ms, respectively [53]. Therefore, the control action time for control resources is fixed and not taken as the decision variable in the decision-making.

4.1.2. Adjustment Amount Constraints

The adjustment amount of each equipment should not exceed its maximum power capacity, such as the maximum active power of HVDC can be increased up to being 1.1 times the rated capacity [54]. Therefore, the emergency control strategies should meet the following constraints:

$$p_i^{\mathrm{DC,min}} - p_i^{\mathrm{DC}} \le \Delta p_i^{\mathrm{DC}} \le p_i^{\mathrm{DC,max}} - p_i^{\mathrm{DC}} \; (i = 1, \cdots, N_{\mathrm{D}}), \tag{6}$$

$$p_k^{\text{load,min}} - p_k^{\text{load}} \leq \Delta p_k^{\text{load}} \leq p_k^{\text{load,max}} - p_k^{\text{load}} \ (k = 1, \cdots, N_{\text{L}}), \tag{7}$$

where p_i^{DC} is the transmission power of HVDC i; $p_i^{\text{DC,max}}$ and $p_i^{\text{DC,min}}$ are the transmission power limits of HVDC i; p_k^{load} is the power of load k; and $p_k^{\text{load,max}}$ and $p_k^{\text{load,min}}$ are the LS amount limits of load k.

Equation (6) is the power adjustment amount limits of HVDC and Equation (7) is the LS limits.

4.1.3. Steady-State Constraints

Based on the indices discussed in Section 3.3, the steady-state constraints are as follows:

$$\Delta f^{\text{s,min}} < \Delta f^{\text{s}}(p) < \Delta f^{\text{s,max}}, \tag{8}$$

$$\Delta V_m^{\text{s,min}} < \Delta V_m^{\text{s}}(p) < \Delta V_m^{\text{s,max}} \ (m = 1, \cdots, N_{\text{B}}), \tag{9}$$

$$S_q^{\text{s}}(p) < S_q^{\text{s,max}} \ (q = 1, \cdots, N_{\text{T}}), \tag{10}$$

where $\Delta f^{\text{s,max}}$ and $\Delta f^{\text{s,min}}$ are the steady-state frequency deviation limits of the system; N_{B} is the total number of the buses; $\Delta V_m^{\text{s,max}}$ and $\Delta V_m^{\text{s,min}}$ are the upper and lower limits of the steady-state voltage deviation at bus m; N_{T} is the total number of lines; S_q^{s} is the transmission power of line q; and $S_q^{\text{s,max}}$ is the transmission power limit of line q.

Equations (8) and (9) are the steady-state deviation limits of the frequency and voltage. Equation (10) is the transmission power limit of lines.

4.1.4. Transient-State Constraints

The transient variables, such as the transient frequency deviation, transient voltage deviation, and relative power angle, of the generators should meet:

$$\Delta f_g^{\text{d,min}} < \Delta f_g^{\text{d}}(p) < \Delta f_g^{\text{d,max}} \ (g = 1, \cdots, N_{\text{G}}), \tag{11}$$

$$\Delta V_m^{\text{d,min}} < \Delta V_m^{\text{d}}(p) < \Delta V_m^{\text{d,max}} \ (m = 1, \cdots, N_{\text{B}}), \tag{12}$$

$$\Delta \delta_{s,r}(p) < \Delta \delta^{\text{max}} \ (s, r = 1, \cdots, N_{\text{G}}), \tag{13}$$

where N_{G} is the total number of the generators; $\Delta f_g^{\text{d,max}}$ and $\Delta f_g^{\text{d,min}}$ are the upper and lower limits of the transient frequency deviation at generator g; $\Delta V_m^{\text{d,max}}$ and $\Delta V_m^{\text{d,min}}$ are the upper and lower limits of the transient voltage deviation at bus m; $\Delta \delta_{s,r}$ is the power angle difference between generators s and r; and $\Delta \delta^{\text{max}}$ is the maximum power angle difference between any two units during the transient process.

Equations (11) and (12) are the transient deviation limits of the frequency and voltage. Equation (13) is the transient limit of the power angle difference.

4.2. Decision-Making Procedure of the Emergency Control Strategy

As described in Section 4.1.1, the priority and the control action time of the emergency resources are different. In the decision-making process, the resources are optimized in order of priority, which is HVDCs, pumped storages, and interruptible loads. Only when the adjustable amount of the resource with high priority is insufficient to maintain the security and stability will the resource with low priority be adjusted. Therefore, the types of control resources that need to be adjusted should be determined firstly according to the power shortage and the control strategy obtained from the off-line table. Then, those with high priority are adjusted to the maximum adjustment amount, and those with low priority are optimized by solving the decision-making model.

As for the solution method, there are two kinds that can be used to solve the non-linear decision-making problems: One is to transform the non-linear function to a linear function, such as the trajectory sensitivity-based method in [30], and the other is to handle the problems with AI algorithms. In the study, the latter one is adopted, in which the BAS algorithm [40] and TS simulation are combined to obtain the optimal control strategy. Considering that the BAS algorithm may take several iterations during the decision-making process and the influence of the control strategies brought by the steady-state model in TS simulation is relatively small, TS simulation is used to improve the overall efficiency. At the same time, it should be noted that the decision variable corresponding to the pumped storages is an integer variable. In the optimization, it is treated as a continuous variable, and finally rounded to the nearest integer to obtain the decision-making result.

The specific decision-making procedure is as follows and the flowchart is shown in Figure 4.

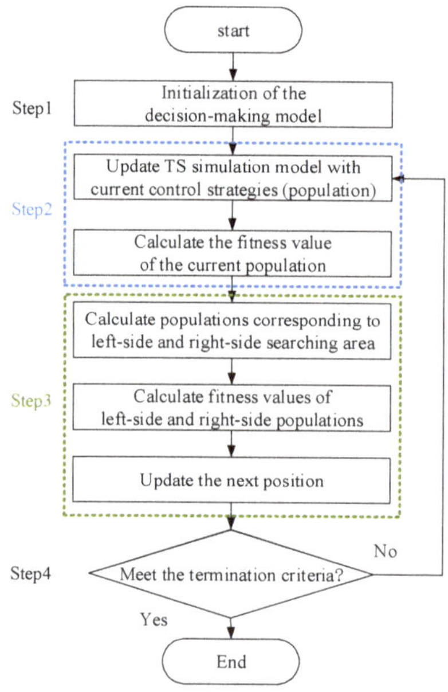

Figure 4. Flowchart of the decision-making procedure.

Step 1: Initialization of the decision-making model.

Determine the types of control resources that need to be adjusted through comparing the adjustable amount of the resources and the power shortage. Obtain the control strategy corresponding to the pre-determined contingency and the current operating condition through an approximate search of the off-line control strategy table. If the control resource types in the control strategy are the same as those determined based on the power shortage, then the control strategy is used as the initial population x; otherwise, if the control resource types in the control strategy differ from those determined based on the power shortage, the control resource types are consistent with those determined based on the power shortage, and the resource with the lowest priority in the control resource types will be optimized, with the initial adjustment amount as 0. Then, initialize the decision-making model with the current operating state data, initial population x, and other solution parameters. The solution parameters include the variable step-size parameter E, the step-size s^p, the distance between left and right populations d_0, and the number of iterations n.

Step 2: Fitness value calculation of the current population.

Update the TS simulation model with the current control strategy, i.e., the current population x. Then, extract the deviations of the frequency and voltage described in Section 3.3 through traversing the simulation results. Finally, calculate the fitness value of the current population based on the fitness value function shown in Equations (1)–(5).

Step 3: Update of the population.

Assume that the beetle forages randomly in any direction, then the direction vector from its right antenna to the left antenna should also be random. Therefore, the optimization problems in $k^{\dim}$ dimensional space can be represented and normalized by a random vector:

$$D = \frac{rands(k^{\dim}, 1)}{\left|rands(k^{\dim}, 1)\right|}, \tag{14}$$

where $k^{\dim}$ is the spatial dimension and $rands()$ is a random function.

To imitate the activities of the beetle's left and right antennae, populations x_l and x_r are defined to represent a population in the left-side and right-side searching areas, respectively:

$$x_l - x_r = d_0 \cdot D, \tag{15}$$

$$x_l = x + d_0 \cdot D/2, \tag{16}$$

$$x_r = x - d_0 \cdot D/2. \tag{17}$$

Then, the fitness values of populations x_l and x_r are calculated based on TS simulation results and Equations (1)–(5), and expressed as f^{left} and f^{right}, respectively.

Finally, the position where the beetle will go next, i.e., the next population, can be determined by comparing the fitness values f^{left} and f^{right} based on Equation (18):

$$x = \begin{cases} x + E \cdot s^{\text{p}} \cdot D \left(f^{\text{left}} < f^{\text{right}}\right) \\ x - E \cdot s^{\text{p}} \cdot D \left(f^{\text{left}} > f^{\text{right}}\right) \end{cases}. \tag{18}$$

The variable step-size parameter E is between 0 and 1, and 0.95 is an acceptable value here.

Step 4: Termination criteria

If the difference between the fitness values of two adjacent populations is less than the threshold value ε or the number of iterations n has reached the maximum value, as shown in Equation (19), then the decision-making is terminated and the new population is considered as the optimal emergency control strategy; otherwise, take the previous population as the input and perform step 2 and step 3 again until Equation (19) is met:

$$f_n - f_{n-1} \leq \varepsilon \text{ or } n \geq n^{\max}, \tag{19}$$

where f_n and f_{n-1} are the fitness values of the nth iteration and $(n-1)$th iteration, respectively; and $n^{\max}$ is the maximum number of iterations.

5. Case Studies

In this section, two actual power systems in China are used as the test systems to verify the proposed scheme.

5.1. Test System 1

The topology of test system 1 is shown in Figure 5. There are 64 equivalent loads, 39 equivalent generators, 101,000 kV buses, 80,500 kV buses, and 3 HVDC lines: ±660 HVDC 1, ±800 HVDC 2, and ±800 HVDC 3. The total capacity of equivalent loads is 59.6 GW, and the transmission power of the HVDC lines are 4, 8, and 8 GW, respectively. That is, the capacity proportion of HVDCs is 33.56% of the equivalent loads.

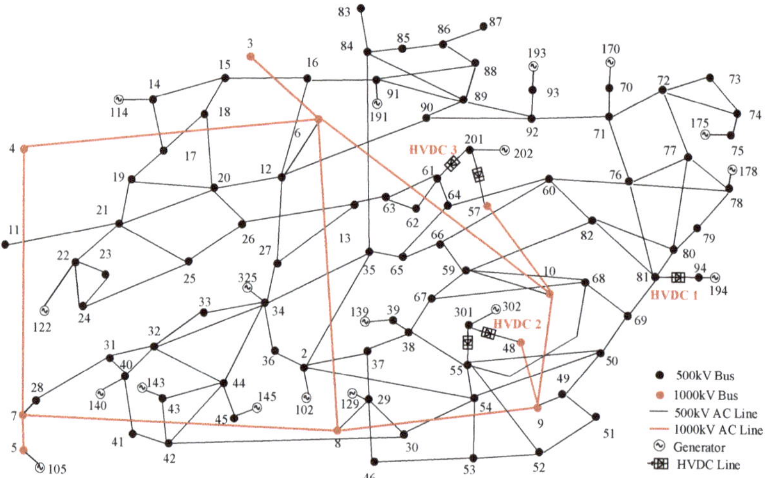

Figure 5. Topology of test system 1.

5.1.1. Construction of the Hybrid Simulation

As discussed in Section 3.2.1, the accuracy of the hybrid simulation is related to the interface buses, so the frequency-domain characteristics analysis is conducted to determine the interface buses.

For the convenience of description, the number of branches in the shortest path between two buses is defined as the electrical distance. For example, the electrical distance between bus 38 and bus 66 is 3. Since the commutation buses are modeled as the internal nodes of the HVDC model in PSS/E, the buses with an electrical distance of 2, 3, 4, and 5 from the commutation buses are taken as the interfaces to construct the hybrid simulation models, respectively. According to Section 3.2.1, the impedance-frequency characteristics at the commutation buses of HVDCs are obtained based on the frequency-domain characteristics analysis. Take HVDC 2 as an example, the positive-sequence impedance-frequency characteristics of the rectifier-side bus 301 and inverter-side bus 55 are shown in Figure 6.

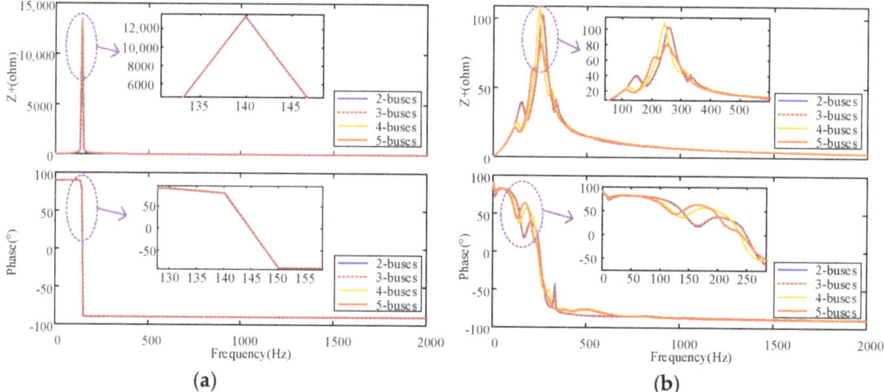

Figure 6. (**a**) Impedance-frequency characteristic of the rectifier-side bus 301; (**b**) Impedance-frequency characteristic of the inverter-side bus 55.

As can be seen in the figures, the differences of the impedance-frequency characteristics at bus 301 under different interface locations are negligible, which may result from the direct connection with generator 302. Additionally, the four waveforms at bus 55 match well when the frequency is lower than 110 Hz and higher than 400 Hz. Although some differences exist under other frequencies, the characteristics under 2 buses away and 3 buses away are very close. Buses with an electrical distance of 2 from the commutation buses are taken as the interface location.

5.1.2. Implementation of Security Assessment and Emergency Control Strategy Decision-Making

As discussed in Section 2, operation failure of protection and reclosing failure caused by a permanent fault are two issues of interest to researchers in recent years. Therefore, they are studied as two scenarios in the study. To verify the accuracy of the proposed EMT-TS hybrid simulation, the PSS/E simulator is adopted as the pure TS simulation tool for comparison.

In the EMT-TS simulation, the limit of the extinction angle for commutation failures determination is 7.2°. In PSS/E, the actual AC voltage criteria of commutation failures for HVDC 1, HVDC 2, and HVDC 3 are 528, 628, and 628 kV while the criteria of blocking events are 0.6 p.u.

- Scenario 1: Operation Failure of Protection

a. Implementation of Security Assessment.

In this case, a three-phase short-circuit fault occurs at line from bus 29 to bus 46 at 1.1 s, and the opening of the circuit breaker fails due to its malfunction. Therefore, the faulted line is finally isolated by tripping circuit breakers of adjacent lines at 1.4 s, which is called failure protection.

Figure 7 shows the corresponding responses of typical interface buses and HVDC 2 in the hybrid simulation and PSS/E. As can be seen from Figure 7a, the waveforms of interface buses match well before the fault occurs. Although there is a slight deviation in the transient process before the fault removal, a similar trend is obtained, which can verify the correctness of the hybrid simulation results. Meanwhile, continuous commutation failures of HVDC 2 are observed in both PSS/E and hybrid simulations during the fault. It should be noted that due to the different modeling methods of HVDC converters, the extinction angle under commutation failures is different in PSS/E and the hybrid simulation. In PSS/E, the extinction angle is set to 90° [45], while in the hybrid simulation, the extinction angle is lower than 7.2° [46]. Therefore, it can be seen from the waveforms of the extinction angle in Figure 7b, in both PSS/E and the hybrid simulation, the intervals between two commutation failures (extinction angle is lower than 7.2° in the hybrid simulation while equals to 90° in PSS/E) are longer than 200 ms, which indicates the occurrence of continuous commutation failures. Nevertheless, HVDC 2 is blocked at 1.4 s in the hybrid simulation while not in PSS/E, which can be seen from the slow restoration of the inverter-side active power in PSS/E. Therefore, it validates that the ETM-TS hybrid simulation proposed in this paper can detect the blocking event while there is a limitation in using pure TS simulation to detect blocking events.

Through traversing the simulation results of scenario 1, it can be found that the steady-state frequency deviation $|\Delta f|$ is 0.24956, which exceeds the threshold of 0.05, and the minimum transient frequency is 49.169, which is lower than the threshold of 49.25 Hz. Therefore, the emergency control strategy should be developed to maintain the security and stability of the receiving-end system.

b. Implementation of Emergency Control Strategy Decision-Making.

Since there is no pumped storage in the provincial power system, only HVDCs and interruptible loads are taken as the control resources. By applying the decision-making method proposed in Section 4.2, the emergency control strategy for the bipolar blocking event of HVDC 2 is to increase the transmission power of the rest HVDC systems by 1.2 GW at 1.5 s and shear a load of 6.16 GW at 1.7 s. The static security indices and dynamic security indices before and after adopting the emergency control strategy are shown in Figure 8. The steady-state and transient frequency indices will exceed the preset range without control, while all static and dynamic indices are within preset ranges with the control strategy obtained by the proposed method.

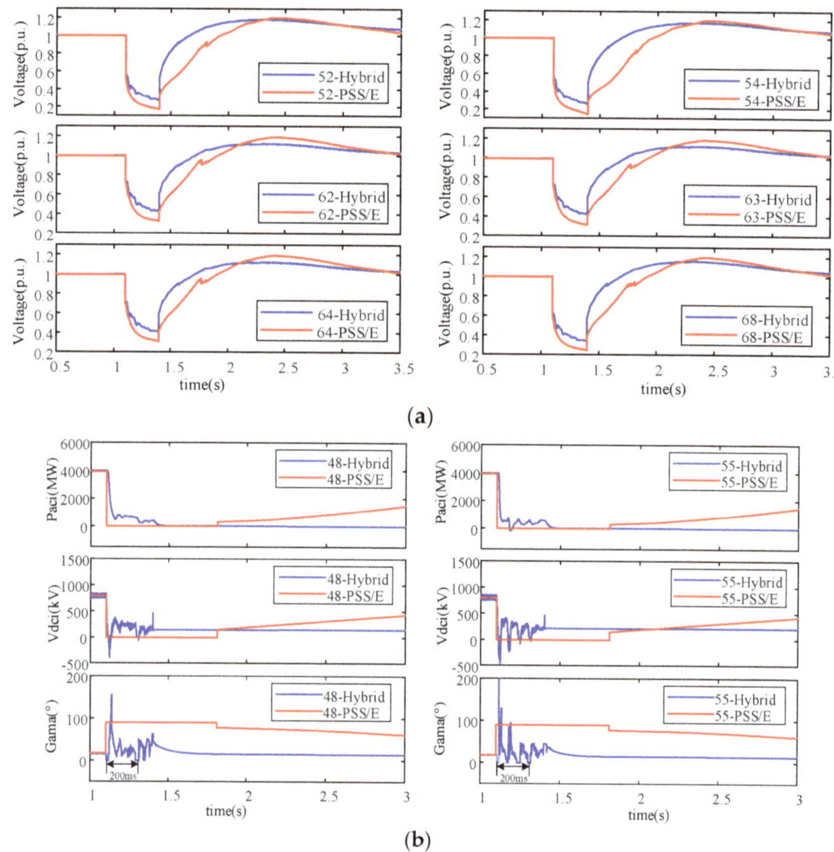

Figure 7. (**a**) Voltage of typical interface buses; (**b**) Active power, dc voltage, and extinction angle at the inverter side of HVDC 2.

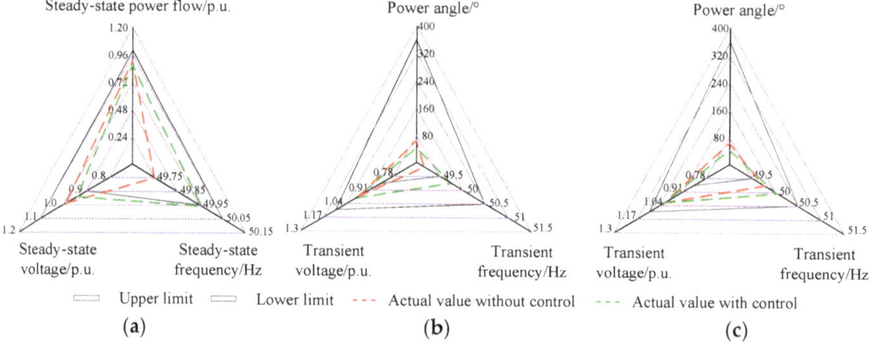

Figure 8. (**a**) Static indices; (**b**) Dynamic indices corresponding to the minimum transient frequency and voltage; (**c**) Dynamic indices corresponding to the maximum transient frequency and voltage.

In order to further verify the control effect of the emergency control strategy, the trajectory sensitivity-based LS scheme proposed in [55] is compared with the proposed scheme in the paper, and the results are shown in Figure 9. The LS ranges of the sensitivity-based scheme are set as (0, 10%) and (0, 14%), respectively. As can be seen from the results, the LS amount under the two ranges are concentrated at the upper or lower limit, and there is significant non-uniformity. The control costs are 6.4763 and 6.4712 GW, respectively. In comparison, the LS amount obtained from the proposed scheme has higher consistency among the entire network, and the local LS is not uniform. Furthermore, the control cost of LS is reduced to 6.1646 GW.

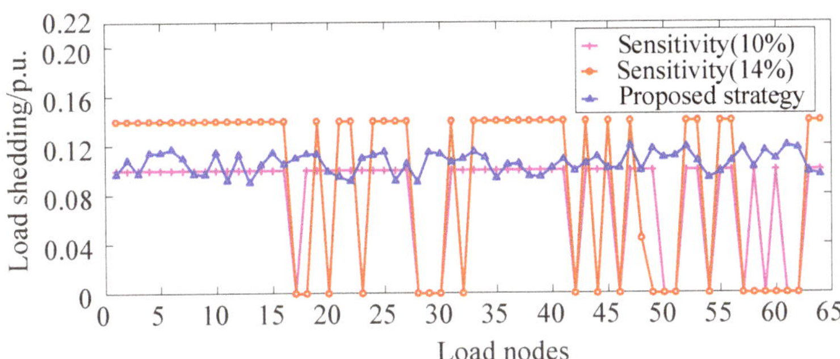

Figure 9. Load shedding amount under different schemes.

- Scenario 2: Reclosing at a Permanent Fault

a. Implementation of Security Assessment.

In this scenario, a three-phase short-circuit fault occurs at the line between bus 65 and 66 at 1.1 s, and the circuit breaker is opened at 1.2 s. In addition, the reclosing of the circuit breaker at 2.2 s fails due to a permanent fault. Therefore, the circuit breaker is reopened at 2.3 s.

The results of typical interface buses, HVDC 2 and HVDC 3, are shown in Figure 10. It can be seen in Figure 10a that the voltage waveforms of interface buses in hybrid simulation and PSS/E before reclosing are close. However, the HVDC systems show different characteristics during the transient process. As can be seen in Figure 10b,c, in the hybrid simulation, continuous commutation failures are observed in HVDC 2 and HVDC 3 due to the unsuccessful reclosing of the breaker, so they are blocked at 2.3 s; while in PSS/E, the active power of the HVDC systems restores slowly after the reopening of the breaker. Obviously, reclosing to a permanent fault does not cause the second commutation failure in PSS/E, which shows the limitation of adopting the AC voltage at the inverter side as the criterion for detecting the commutation failure.

Different from Scenario 1, in addition to the steady-state frequency deviation and the minimum transient frequency, the maximum transient frequency exceeds the threshold. Therefore, the emergency control strategy should be developed.

b. Implementation of Emergency Control Strategy Decision-Making.

As discussed in the above, the permanent fault will cause bipolar blocking events of HVDC 2 and HVDC 3, leading to a power loss of 16 GW. Through applying the decision-making method proposed in Section 4.2, the emergency control strategy is to increase the transmission power of the rest HVDC systems by 0.4 GW at 2.4 s and shear a load of 15.3 GW at 2.6 s. The static and dynamic indices before and after adopting the emergency control strategy are shown in Figure 11. The steady-state and transient frequency indices will exceed the preset ranges without control, while all static and dynamic indices are within preset ranges with the control strategy obtained by the proposed method.

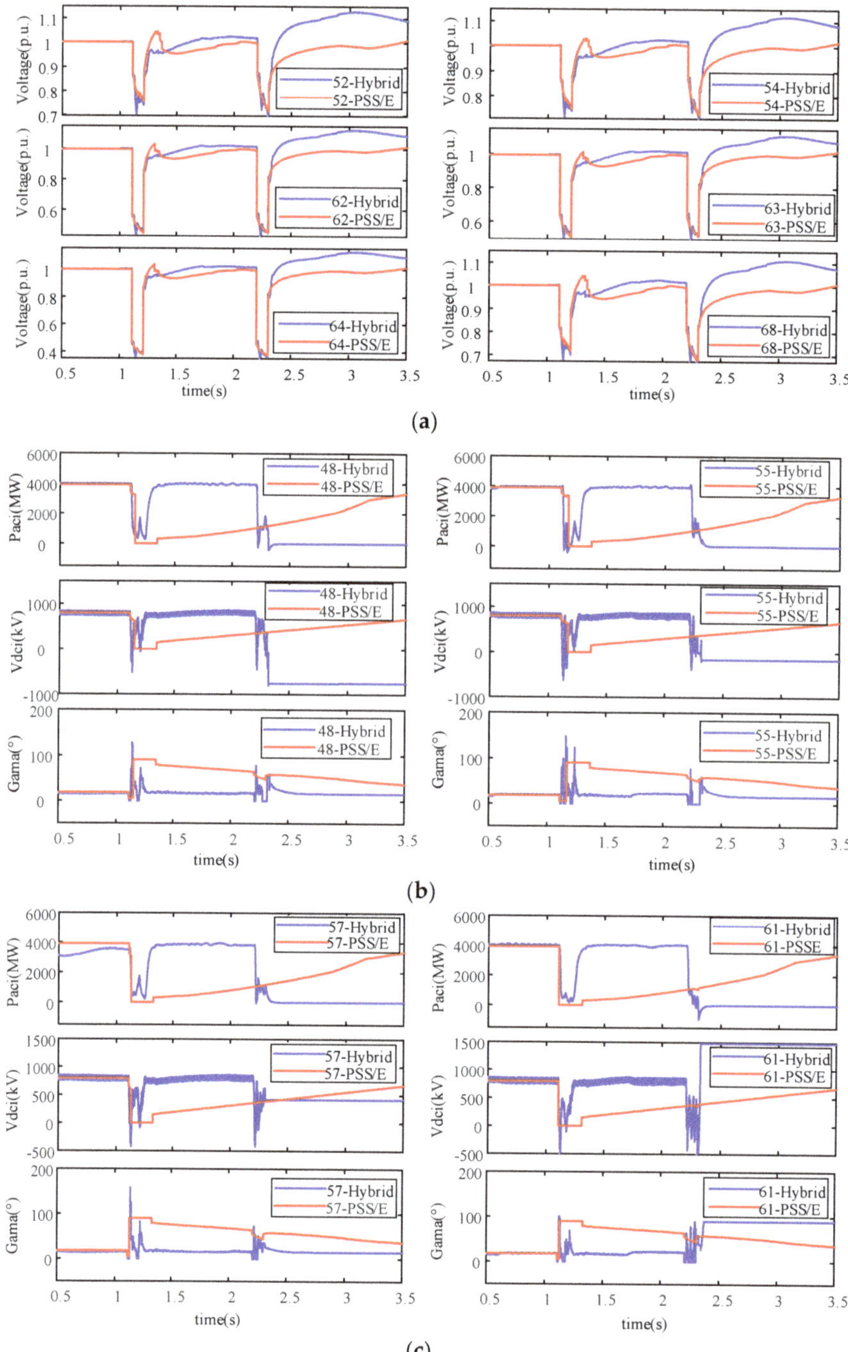

Figure 10. (**a**) Voltage of interface buses; (**b**) Active power, dc voltage and extinction angle at the inverter side of HVDC 2; (**c**) Active power, dc voltage, and extinction angle at the inverter side of HVDC 3.

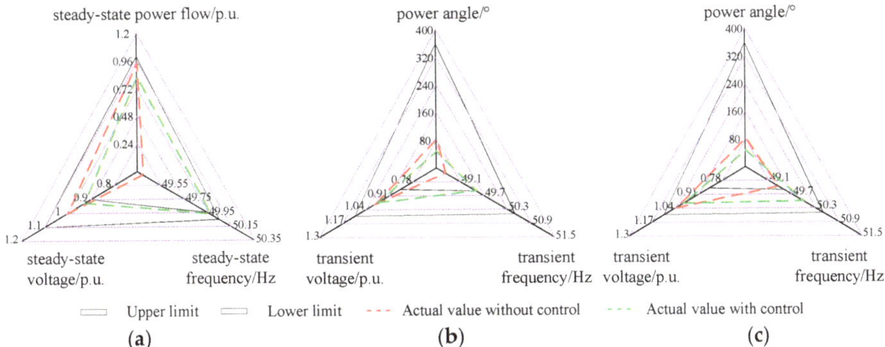

Figure 11. (**a**) Static indices; (**b**) Dynamic indices corresponding to the minimum transient frequency and voltage; (**c**) Dynamic indices corresponding to the maximum transient frequency and voltage.

5.2. Test System 2

Test system 2 is divided into three regional grids by 8 HVDC lines, as shown in Figure 12. There are 271 buses and 296 AC transmission lines. The total capacity of the generators and loads are 27,550 and 26,878 MW, respectively. For the HVDCs, the rated voltage is ±800 kV and the transmission power is 800 MW, respectively. The specific information of regions is shown in Table 3.

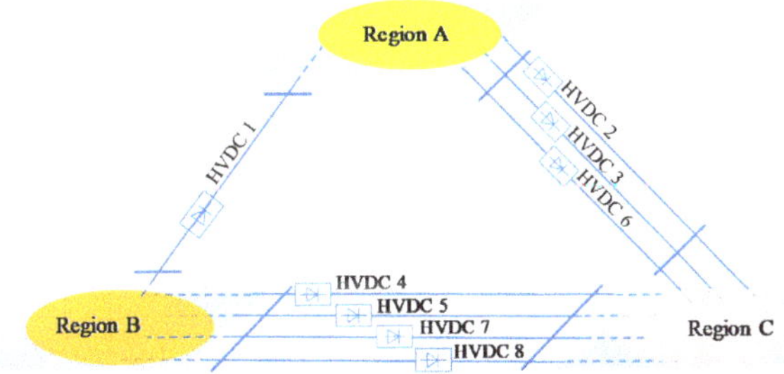

Figure 12. Topology of test system 2.

Table 3. Specific information of regions.

Regions	Capacity of Generators (MW)	Capacity of Loads (MW)	Sending HVDC (MW)	Feeding HVDC (MW)
Region A	8100	6363	3200	800
Region B	10,000	5653	4000	0
Region C	9450	14,862	0	5600

Due to the space limitations, only the results of the emergency control strategy are presented. Assume that HVDC 3 is blocked at 1.4 s under the scenario of operation failure of protection. The steady- state frequency and the transient frequency of region A and C exceed the threshold. Therefore, the emergency control strategy is developed based on the decision-making method proposed in Section 4.2.

The emergency control strategy for the bipolar blocking event of HVDC 3 is to increase the transmission power of the rest HVDC systems between region A and C by 160 MW (HVDC 2 and HVDC 6), HVDC systems between region B and C by 320 MW (HVDC 4, HVDC 5, HVDC 7, and HVDC 8), and decrease the transmission power of HVDC 1 by 320 MW at 1.5 s. Then, a generator of 600 MW in region A is sheared at 1.6 s and a load of 500 MW in region C at 1.7 s. The static and dynamic security indices of the three regions before and after adopting the emergency control strategy are shown in Figures 13–15, respectively. The steady-state and transient frequency indices of region A and C finally meet the preset ranges with the control strategy obtained by the proposed method.

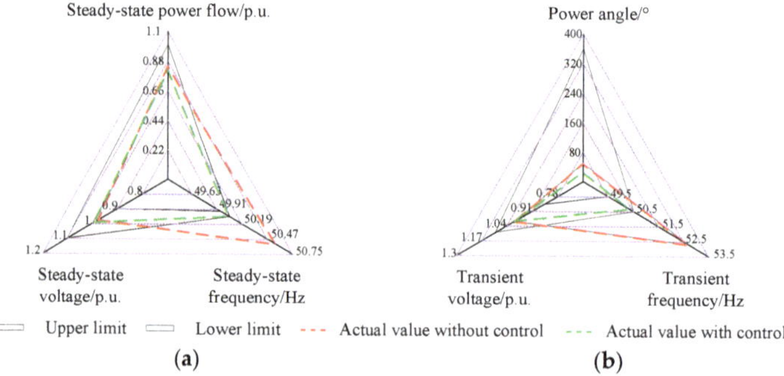

Figure 13. Static and dynamic security indices of region A. (**a**) Static indices; (**b**) Dynamic indices corresponding to the maximum transient frequency and voltage.

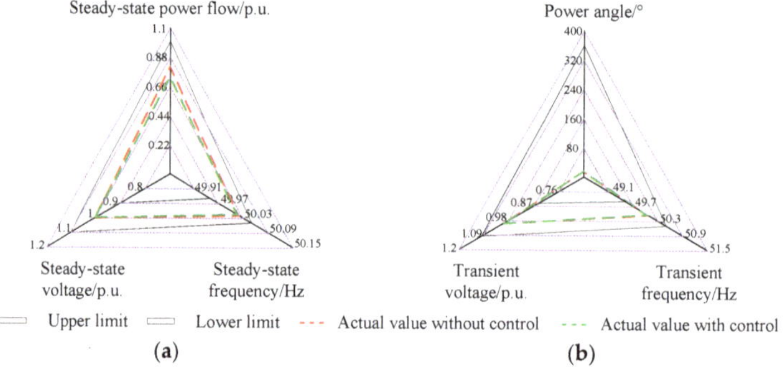

Figure 14. Static and dynamic security indices of region B. (**a**) Static indices; (**b**) Dynamic indices corresponding to the maximum transient frequency and voltage.

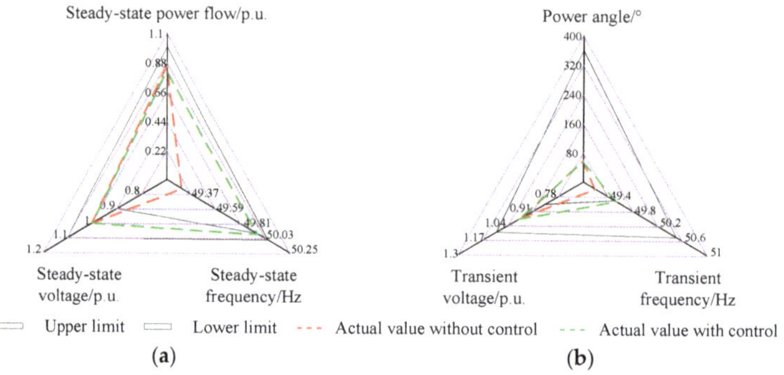

Figure 15. Static and dynamic security indices of region C. (**a**) Static indices; (**b**) Dynamic indices corresponding to the minimum transient frequency and voltage.

6. Conclusions

This paper proposes an on-line pre-decision-making scheme, including security assessment and an emergency control strategy decision-making, for power systems with multi-infeed HVDCs. The security assessment method is based on the EMT-TS hybrid simulation, and can generate accurate assessment results while maintaining the high computational efficiency. The emergency control strategy decision-making method can make full use of HVDCs, pumped storages, and interruptible loads to maintain the security and stability of receiving-end systems. The case studies showed that the proposed scheme is reliable. In addition, the results also indicate that it is essential to describe interactions between AC and DC subsystems in security assessment to identify HVDC-related security and stability issues.

In future work, the dynamic average-value modeling method can be used in HVDC modeling to further improve the computational efficiency. In addition, more attention will be paid to the detailed design of the emergency control strategy, such as the coordination of HVDC controllers and the classification of interruptible loads.

Author Contributions: All authors contributed to the research in the paper. Conceptualization, Y.W. and Y.X.; Data curation, Z.S.; Formal analysis, Q.Z. and Z.S.; Funding acquisition, J.H.; Investigation, Q.Z.; Methodology, Q.Z. and Z.S.; Project administration, Y.W., J.H. and Y.X.; Resources, M.L.; Software, Q.Z. and Z.S.; Supervision, J.H., Y.X. and M.L.; Validation, Y.W.; Visualization, M.L.; Writing—original draft, Q.Z. and Z.S.; Writing—review & editing, Y.W. and Y.X. and all the authors have read and approved the final manuscript. All authors have read and agreed to the published version of the manuscript.

Funding: This work was funded by the Fundamental Research Funds for the Central Universities, Grant No. 2018YJS155, and the National Key R&D Program of China, Grant No. 2017YFB0902600.

Conflicts of Interest: The authors declare no conflict of interest.

References

1. Pedroso, F.R.V.A.; Bassini, M.T.; Horita, M.A.B.; Jardini, J.A.; Graham, J.F.; Liu, G. HVDC multi-infeed analysis of the Brazilian transmission system and possible mitigation methods. *CSEE J. Power Energy Syst.* **2018**, *4*, 487–494. [CrossRef]
2. Li, X.; Li, Y.; Liu, L.; Wang, W.; Li, Y.; Cao, Y. Latin hypercube sampling method for location selection of multi-infeed HVDC system terminal. *Energies* **2020**, *13*, 1646. [CrossRef]
3. Li, H.C.; Yuan, Y.B.; Bian, Z.D.; Xu, W. Analysis of frequency emergency control characteristics of UHV AC/DC large receiving end power grid. *J. Eng.* **2017**, *36*, 27–31. [CrossRef]
4. Zhou, B.R.; Rao, H.; Wu, W.; Wang, T.; Hong, C.; Huang, D.Q.; Yao, W.F.; Su, X.R.; Mao, T. Principle and application of asynchronous operation of china southern power grid. *IEEE J. Emerg. Sel. Top. Power Electron.* **2018**, *6*, 1032–1040. [CrossRef]
5. Aik, D.L.H.; Andersson, G. Impact of renewable energy sources on steady-state stability of weak AC/DC systems. *CSEE J. Power Energy Syst.* **2017**, *3*, 419–430. [CrossRef]
6. Zhang, F.; Xin, H.H.; Wu, D.; Wang, Z.; Gan, D.Q. Assessing strength of multi-infeed LCC-HVDC systems using generalized short-circuit ratio. *IEEE Trans. Power Syst.* **2019**, *34*, 467–480. [CrossRef]
7. Shao, Y.; Tang, Y. Fast Evaluation of commutation failure risk in multi-infeed HVDC systems. *IEEE Trans. Power Syst.* **2017**, *33*, 646–653. [CrossRef]
8. Wu, P.; Zhao, B.; Guo, J.B.; Bu, G.Q. Coordinated control method of interconnected grid to solve cascading faults of UHVDC valve groups. *Power Syst. Technol.* **2017**, *41*, 1460–1467.
9. Luo, J.; Gong, Y.; Li, H.C. Online emergency control and corrective control coordination strategy for UHVDC blocking faults. In Proceedings of the International Conference on Electronics Technology, Chengdu, China, 23–27 May 2018; pp. 5–10.
10. Tian, F.; Zhang, X.; Yu, Z.H.; Qiu, W.J.; Shi, D.Y.; Qiu, J.; Liu, M.; Li, Y.L.; Zhou, X.X. On-line decision-making and control of power system stability based on super-real-time simulation. *CSEE J. Power Energy Syst.* **2016**, *2*, 95–103. [CrossRef]
11. Bao, Y.H.; Xu, T.S.; Zhou, H.; Ren, X.C.; Lou, B.L.; Wu, F. An online pre-decision method for security and stability emergency regulation. *Electr. Power* **2019**, *52*, 91–97.
12. Lin, L.; Zhang, K.; Sun, F.; Li, Q.J. Comparison and conversion of dynamic models of speed governor for transient stability analysis in BPA and PSS/E. In Proceedings of the Asia-Pacific Power and Energy Engineering Conference, Chengdu, China, 28–31 March 2010; pp. 1–4.

13. Cui, Y.; Feng, N.; Feng, Y.Y.; Yu, Y.H.; Wang, W.H. Evaluation of security and stability risk of AC/DC grid under extreme contingencies. In Proceedings of the Asia-Pacific Power and Energy Engineering Conference, Xi'an, China, 25–28 October 2016; pp. 1284–1290.
14. Li, H.; Diao, R.; Zhang, X.; Lin, X.; Lu, X.; Shi, D.; Wang, Z.W.; Wang, L. An integrated online dynamic security assessment system for improved situational awareness and economic operation. *IEEE Access* **2019**, *7*, 162571–162582. [CrossRef]
15. Wang, Y.; Li, K.; Zhang, G.H.; Su, J.J.; Xie, J.P.; Li, J.L. Online dynamic voltage stability assessment method of AC/DC power systems. *J. Eng.* **2019**, *16*, 1356–1360. [CrossRef]
16. Zadkhast, S.; Jatskevich, J.; Vaahedi, E. A multi-decomposition approach for accelerated time-domain simulation of transient stability problems. *IEEE Trans. Power Syst.* **2015**, *30*, 2301–2311. [CrossRef]
17. Kim, S.; Overbye, T.J. Optimal subinterval selection approach for power system transient stability simulation. *Energies* **2015**, *8*, 11871–11882. [CrossRef]
18. Zhan, F.J.; Du, Z.B.; Zhao, F.; Zhang, Y. Analysis of the transient voltage stability of AC/DC system based on numerical energy function method. In Proceedings of the Asia-Pacific Power and Energy Engineering Conference, Xi'an, China, 25–28 October 2016; pp. 2204–2208.
19. Chen, X.; Xu, Y.; He, J.H. Transient stability analysis of AC-DC hybrid power systems using direct method. In Proceedings of the 2nd International Electrical and Energy Conference (CIEEC), Beijing, China, 4–6 November 2018; pp. 76–81.
20. Xu, Y.; Dong, Z.Y.; Zhang, R.; Xue, Y.; Hill, D.J. A decomposition-based practical approach to transient stability-constrained unit commitment. *IEEE Trans. Power Syst.* **2015**, *30*, 1455–1464. [CrossRef]
21. Bhui, P.; Senroy, N. Real-time prediction and control of transient stability using transient energy function. *IEEE Trans. Power Syst.* **2017**, *32*, 923–934. [CrossRef]
22. Xu, T.; Li, G.; Yu, Z.; Zhang, J.; Wang, L.; Li, Y.; Zhang, Y.; Zhuang, K.; Luo, J.; Li, D. Design and Application of emergency coordination control system for multi-infeed HVDC receiving-end system coping with frequency stability problem. *Autom. Electr. Power Syst.* **2017**, *41*, 98–104.
23. Yin, J.J. Research on load friendly interactive technology for safe operation of UHV interconnected power grid. *Proc. CSEE* **2016**, *36*, 5715–5723.
24. Ren, C.; Xu, Y. A fully data-driven method based on generative adversarial networks for power system dynamic security assessment with missing data. *IEEE Trans. Power Syst.* **2019**, *34*, 5044–5052. [CrossRef]
25. Zhang, R.Y.; Wu, J.Y.; Xu, Y.; Li, B.Q.; Shao, M.Y. A hierarchical self-adaptive method for post-disturbance transient stability assessment of power systems using an integrated CNN-based ensemble classifier. *Energies* **2019**, *12*, 3217. [CrossRef]
26. Wang, H.; Chen, Q.; Zhang, B. Transient stability assessment combined model framework based on cost-sensitive method. *IET Gener. Transm. Distrib.* **2020**, *14*, 2256–2262. [CrossRef]
27. Szechtman, M.; Ximenes, M.J.; Saavedra, A.R. Comparative analysis of stability and electromagnetic transient studies for HVDC multi-infeed systems. *CSEE J. Power Energy Syst.* **2017**, *3*, 253–259. [CrossRef]
28. Shekari, T.; Aminifar, F.; Sanaye-Pasand, M. An analytical adaptive load shedding scheme against severe combinational disturbances. *IEEE Trans. Power Syst.* **2016**, *31*, 4135–4143. [CrossRef]
29. Sun, D.; Zhou, H.; Ju, P.; Zhou, R.; Su, D.; Xu, C. Optimization method for emergency load control of receiving-end system considering coordination of economy and voltage stability. *Autom. Electr. Power Syst.* **2017**, *41*, 106–112.
30. Xu, X.; Zhang, H.; Li, C.; Liu, Y.; Li, W.; Terzija, V. Optimization of event-driven emergency load-shedding considering transient security and stability constraints. *IEEE Trans. Power Syst.* **2017**, *32*, 2581–2592. [CrossRef]
31. Zhang, C.L.; Chu, X.D.; Zhang, B.; Ma, L.L.; Li, X.; Wang, X.B.; Wang, L.; Wu, C. A coordinated DC power support strategy for multi-infeed HVDC systems. *Energies* **2018**, *11*, 1637. [CrossRef]
32. Zhang, N.Y.; Zhou, Q.; Hu, H.M. Minimum frequency and voltage stability constrained unit commitment for AC/DC transmission systems. *Appl. Sci.* **2019**, *9*, 3412. [CrossRef]
33. Song, Y.Z.; Meng, J. Optimization strategy of under-frequency load shedding for pumped storage units in power system. *Meas. Control Technol.* **2017**, *36*, 52–56.
34. Li, H.; Yuan, Y.; Zhang, X.; Su, C. The frequency emergency control characteristic analysis for UHV AC/DC large receiving end power grid. *Electr. Power Eng. Technol.* **2017**, *36*, 27–31. [CrossRef]

35. Dong, X.J.; Luo, J.B.; Li, X.M.; Li, B.; Wan, F.; Xue, F.; Wang, Z.; Li, Z. Research and application of frequency emergency coordination and control technology in hybrid AC/DC power grids. *Power Syst. Prot. Control* **2018**, *46*, 59–66.
36. Yuan, S.; Chen, D.Z.; Luo, Y.Z.; Ren, J.; Song, Y.; Jia, L.; Zi, P.; Wang, Q.; Li, Z.; Zhang, C.; et al. Stability characteristics and coordinated control measures of multi-resource for DC blocking fault impacting weak AC channel. *Electr. Power Autom. Equip.* **2018**, *38*, 203–210.
37. Shu, Y.B.; Tang, Y.; Sun, H.D. Research on power system security and stability standards. *Proc. CSEE* **2013**, *33*, 1–8.
38. Jing, L.M.; Wang, B.; Dong, X.Z. Review of consecutive commutation failure research for HVDC transmission system. *Electr. Power Autom. Equip.* **2019**, *39*, 116–123.
39. Pipelzadeh, Y.; Moreno, R.; Chaudhuri, B.; Strbac, G.; Green, T.C. Corrective control with transient assistive measures: Value assessment for Great Britain transmission system. *IEEE Trans. Power Syst.* **2017**, *32*, 1638–1650.
40. Jiang, X.; Li, S. BAS: Beetle antennae search algorithm for optimization problems. *Int. J. Robot. Control* **2018**, *1*, 1–5. [CrossRef]
41. Sauer, P.W.; Tomsovic, K.L.; Vittal, V. *Power System Stability and Control*, 2nd ed.; CRC Press: Boca Raton, FL, USA, 1985; Chapter 15.
42. Thio, C.V.; Davies, J.B. Commutation failures in HVDC transmission systems. *IEEE Trans. Power Deliv.* **1996**, *11*, 946–957. [CrossRef]
43. Huang, Y.; Xu, Z.; He, H. HVDC Models of PSS/E and their applicability in simulations. *Power Syst. Technol.* **2004**, *28*, 25–29.
44. Li, W.; Niu, S.B.; Lv, Y.Z.; Ke, X.Y.; Huo, C.; Yu, Y.; Xu, Z.; Wen, Y.; Liu, N. HVDC electromechanical transient model and comparison with HVDC electromagnetic transient model. *High Volt. Appar.* **2018**, *54*, 155–160.
45. PSS/E-High-Performance Transmission Planning and Analysis Software. Available online: https://new.siemens.com/global/en/products/energy/services/transmission-distribution-smart-grid/consulting-and-planning/pss-software/pss-e.html (accessed on 10 May 2019).
46. PSCAD. Available online: https://hvdc.ca/pscad/ (accessed on 5 March 2019).
47. E-Tran. Available online: http://www.electranix.com/software/ (accessed on 15 May 2019).
48. Heffernan, M.D.; Tunrer, K.S.; Arrillaga, J.; Arnold, C.P. Computation of AC-DC system disturbances-Parts I: Interactive coordination of generator and convertor transient models. *IEEE Trans. Power Appar. Syst.* **1981**, *100*, 4341–4348. [CrossRef]
49. Tuner, K.S.; Heffernan, M.D.; Arnold, C.P.; Arrillaga, J. Computation of AC-DC system disturbances-Parts II: Derivation of power frequency variables from convertor transient response. *IEEE Trans. Power Appar. Syst.* **1981**, *100*, 4349–4355. [CrossRef]
50. Reeve, J.; Adapa, R. A new approach to dynamic analysis of ac networks incorporating detailed modeling of DC systems. Part I: Principles and implementation. *IEEE Trans. Power Deliv.* **1988**, *3*, 2005–2011. [CrossRef]
51. Zhang, H.; Li, R.; Song, Y.T.; Liu, W.Z. Research on the interface algorithm of power system electromechanical-electromagnetic hybrid simulation. In Proceedings of the 5th International Conference on Electric Utility Deregulation and Restructuring and Power Technologies (DRPT), Changsha, China, 26–29 November 2015; pp. 296–300.
52. Shi, Z.; Xu, Y.; Wu, X.Y.; He, J.H.; Yang, R.; Yang, M. Assessment of system protection strategy and aided decision scheme for AC/DC hybrid power systems. *Electr. Power Autom. Equip.* **2020**, *40*, 25–31.
53. Chen, G.P.; Li, M.J.; Xu, T. System protection and its key technologies of UHV AC and DC power grid. *Autom. Electr. Power Syst.* **2018**, *42*, 8–16.
54. Li, G.; Liang, J.; Ma, F.; Ugalde-Loo, C.E.; Liang, H.F. Analysis of single-phase-to-ground faults at the valve-side of HB-MMCs in HVDC systems. *IEEE Trans. Ind. Electron.* **2019**, *66*, 2444–2453. [CrossRef]
55. Xu, X.; Zhang, H.X.; Li, C.G.; Liu, Y.K.; Li, W. Emergency load shedding optimization algorithm based on trajectory sensitivity. *Autom. Electr. Power Syst.* **2016**, *40*, 143–148.

MDPI
St. Alban-Anlage 66
4052 Basel
Switzerland
Tel. +41 61 683 77 34
Fax +41 61 302 89 18
www.mdpi.com

Energies Editorial Office
E-mail: energies@mdpi.com
www.mdpi.com/journal/energies

www.ingramcontent.com/pod-product-compliance
Lightning Source LLC
LaVergne TN
LVHW070615170726
843515LV00004B/87

* 9 7 8 3 0 3 6 5 0 3 2 0 2 *